HONING THE KEYS TO THE CITY

Refining the United States Marine Corps Reconnaissance Force for Urban Ground Combat Operations

Russell W. Glenn

Jamison Jo Medby

Scott Gerwehr

Fred Gellert

ndrew O'Donnell

Prepared for the United States Marine Corps

NATIONAL DEFENSE RESEARCH INSTITUTE

RAND

Authorized for public release; distribution unlimited

The research described in this report was sponsored by the United States Marine Corps. The research was conducted in RAND's National Defense Research Institute, a federally funded research and development center supported by the Office of the Secretary of Defense, the Joint Staff, the unified commands, and the defense agencies under Contract DASW01-01-C-0004.

Library of Congress Cataloging-in-Publication Data

Honing the Keys to the city : refining the United States Marine Corps reconnaissance force for urban ground combat operations / Russell W. Glenn ... [et al.].
p. cm.
"MR-1628."
Includes bibliographical references.
ISBN 0-8330-3311-5 (pbk.)
1. United States. Marine Corps. 2. Military reconnaissance. 3. Urban warfare
I. Glenn, Russell W.

VE23.H59 2003
355.4'26'0973—dc21

2002155029

RAND is a nonprofit institution that helps improve policy and decisionmaking through research and analysis. RAND® is a registered trademark. RAND's publications do not necessarily reflect the opinions or policies of its research sponsors.

Cover artwork by Priscilla B. Glenn
Cover design by R. Rodney Sato

© Copyright 2003 RAND

All rights reserved. No part of this book may be reproduced in any form by any electronic or mechanical means (including photocopying, recording, or information storage and retrieval) without permission in writing from RAND.

Published 2003 by RAND
1700 Main Street, P.O. Box 2138, Santa Monica, CA 90407-2138
1200 South Hayes Street, Arlington, VA 22202-5050
201 North Craig Street, Suite 202, Pittsburgh, PA 15213-1516
RAND URL: http://www.rand.org/
To order RAND documents or to obtain additional information, contact Distribution Services: Telephone: (310) 451-7002;
Fax: (310) 451-6915; Email: order@rand.org

PREFACE

The penalty for undertaking urban combat operations without first performing reconnaissance has historically proven very costly on more than one occasion. Yet reconnaissance is considerably more difficult in villages, towns, and cities than in open terrain. The many buildings and other structures can provide cover and concealment for large numbers of a foe's vehicles, personnel, and supplies. Unlike when these assets are hidden in more open ground under foliage or camouflage nets, overhead systems often cannot penetrate the concealment in urban areas. The acquisition of trustworthy and timely combat intelligence must therefore rely on units trained and equipped to conduct ground combat reconnaissance. Urban areas also present special challenges to these men. Undetected movement is difficult in an environment dense with noncombatants and, possibly, enemy. Noise ricochets off hard surfaces so that even a minor slip can compromise a unit's location. Structures and infrastructure block or otherwise disrupt communications. The sum of challenges is considerably greater than the doctrinal, training, and equipment solutions immediately at hand. The purpose of this study is to narrow that unfortunate gap.

This research was sponsored by the U.S. Marine Corps Warfighting Laboratory (MCWL) and was conducted in the International Security and Defense Policy Center of RAND's National Defense Research Institute (NDRI). NDRI is a federally funded research and development center sponsored by the Secretary of Defense and Joint Staff. This report will be of interest to individuals in the governmental and commercial sector whose responsibilities include doctrine, policy design, funding, planning, preparation, or the development of technologies in support of civil or military operations involving urban environments in both the immediate future and the longer term.

CONTENTS

TABLES

SUMMARY

The following analysis was undertaken at the request of the Commanding General, Marine Corps Warfighting Laboratory (MCWL), Quantico, Virginia. The project had three primary components:

- Identify service shortfalls in the area of urban combat ground reconnaissance.
- Evaluate experimental work being conducted by the MCWL in the above area.
- Provide input to assist in the creation of tactics, techniques, and procedures (TTP) for the subject area.

The focal period for the analysis is the immediate future, out to approximately five years from the present. The level of concern is tactical as opposed to operational or strategic, though the three are continuously interdependent, and there will thus be operational and strategic implications of the following discussion. While technological considerations were to be a part of the final product, they were not to dominate.

This report incorporates research and analysis in support of the first and third tasks identified above. The research and analysis involving the second will appear as a separate document.

The methodology employed involved literature searches of pertinent English language publications, including but by no means limited to doctrinal manuals for the U.S. Marine Corps (USMC), U.S. Army, and British Army. An extensive interview program complemented these investigations of written sources. A common characteristic among most of those to whom the authors spoke was either operational urban reconnaissance experience or assignments in Marine Corps reconnaissance units.

Little exists in the way of written guidance for those undertaking these missions. Training experience is very limited. Fortunately, lessons can be drawn from several events, including combat in 1968 Hue, 1973 Suez City, and Grozny at the recent turn of the century. British operations in Northern Ireland, despite their being characterized by stability missions, also provided significant insights. The amount of thought given to the subject by those serving in Marine reconnaissance units was notable. None felt himself an expert in the field of urban combat ground reconnaissance, yet the quality of responses during interviews reflected that the officers and noncommissioned officers confronted with the potential of commitment to these contingencies were actively debating the issues among themselves. Interviewers heard many of the same shortcomings whether speaking to Marines at Camp Pendleton, California, or men at Camp Lejeune, North Carolina. Their recommendations found common ground with those forwarded by men who have led men in urban combat.

That urban reconnaissance will be an increasingly needed skill is evident to any who look at the world's population growth trends and recent U.S. military commitments. That urban operations' demands differ from many confronted on other terrain is evident with but a cursory look. First, the density of people—friendly forces of several nations, enemy personnel, and noncombatants—is greater than is the norm elsewhere. Cultural awareness is always desirable. In urban areas, it will be essential. The information provided by reconnaissance personnel cannot be properly interpreted in ignorance of local social mores. The opportunity for compromise of teams as they move about this terrain or even after occupation of well-selected hides is greater because more people occupy less space. Much interconnectedness can be found within villages, towns, and cities. As the size and complexity of the urban area increases, difficulty in understanding the inanimate physical and social infrastructures grows dramatically. For example, power distribution, transportation, and communications systems can be enormous, while medical care, religious influences, and power relationships take on a new importance for the combatant. Various military activities will also interact in ways or to a degree not often confronted on other terrain. Regular and covert forces may find themselves occupying the same terrain. Coordination between responsible headquarters will be essential to minimize fratricide. The task is more difficult if the various elements are from different services or nations.

However, the density of participants has its advantages. More U.S. and multinational coalition personnel mean that more potential sources of information are in a given area. Private volunteer organizations and nongovernmental organizations (PVOs and NGOs) similarly offer means of better understanding the environment, in particular how noncombatant behavior might be influenced to reduce the dangers of inadvertent casualties in that group. This is no small matter. Roughly 100,000 Filipinos lost their lives during the fight to retake Manila during World War II.

These difficulties challenge all Marine Corps reconnaissance units. With a dearth of written doctrine, individual Marines question how they will infiltrate, exfiltrate, and evacuate casualties during combat missions. They ask how they are to communicate information in an environment where buildings block radio transmissions to such an extent that headquarters only five kilometers distant may not be in contact. As training sites lack much in the way of what actual developing nations' cities present, these men question whether equipment effective elsewhere will be reliable when employed from asphalt and concrete rather than dirt and rock. Shortfalls are extensive—they encompass every aspect of Marine Corps operations. Finding solutions under fire is the least-preferred method of determining the TTP that will bring success.

The complexity of potential solutions matches that of the challenges. An effort to lend some coherence to the discussion of how to address urban reconnaissance led to the formation of four primary themes:

- The urban environment demands almost constant creative adaptation. Its inherent character, compression of space, and related close proximity of participating parties necessitates rapid adjustments in reaction to adversary behavior or to influence that behavior to favor friendly force objectives.
- Tactical ground reconnaissance is a system of systems within a system.

- Urban operations impose extraordinary leadership, training, task organization, and personnel management demands.
- The urban environment makes special demands on equipment and technology. Testing in rural environments does not constitute testing for urban operations.

The nature of urban competition is such that adaptation can occur more quickly than in environments with lesser densities and slower and fewer means of communicating. Therefore, Marine solutions must have an inherent flexibility. The successful tactic of yesterday will be adroitly countered today. Less important than finding the optimum tactic, technique, or procedure is the creation of individual, groups, or families of TTP that can be molded to meet specific situational needs. These can then be used in combination and various sequences. An obvious requirement also exists for command and control structures that permit both real-time and predictive adaptation.

Urban combat ground reconnaissance's status at the tactical level as a system within a larger system of reconnaissance activities is readily apparent. Consistently viewing it as such during analysis aids in understanding the extent to which any TTP must be developed in the service of a much larger information-collection and intelligence system. Tactical ground reconnaissance is essential because it provides input to this larger system that either is otherwise unattainable or provides essential redundancy. TTP developments that ignore this larger perspective are of little value. Similarly, the capabilities that together constitute Marine Corps tactical ground reconnaissance must function together for the good of the whole. One aspect of this complementary interaction is the traditional view of Surveillance and Target Acquisition (STA), divisional reconnaissance, and force reconnaissance assets as being the close-in, interim distance, and deep assets, respectively. If one accepts that all three of these elements have a role to play in a Marine Corps reconnaissance system (and not everyone shares that sentiment), then it lends perspective to two long-standing and heated debates: (1) the role of STA teams as primarily reconnaissance versus shooter capabilities and (2) force reconnaissance as fundamentally a direct action organization.

A thorough consideration of Marine Corps urban reconnaissance requirements has fundamental organizational structure, leadership, and training implications. Long-standing assumptions regarding team size are a subject of considerable debate among those in reconnaissance battalions and force reconnaissance units. The difficulty of succeeding during urban infiltration drives some to a belief that smaller teams should be the standard—that one or two men rather than four or more is appropriate. Others recognize the undoubted advantages in stealth that fewer numbers bring, but they are concerned that sacrifices in load-carrying capability, security, and ability to defend the team outweigh the benefits. (Some believed one-man missions were desirable; others felt as many as 12 Marines should be the norm for a reconnaissance team.) What was ultimately apparent was that team size will be mission dependent. That much does not distinguish urban reconnaissance from those on other terrains. It is the greater frequency with which missions could dictate use of other than six-man teams that delineates actions in built-up areas, just as this frequency will influence command and control, munitions and weapons carried, and the time needed for accomplishing assigned tasks.

Density of forces and noncombatants, line-of-sight interruption, reflection of sound off hard surfaces, and other urban characteristics suggest that reconnaissance units operating in urban areas will sometimes require materiel different from that used elsewhere. The manner in which commanders employ complementary systems, such as ground reconnaissance and unmanned aerial vehicles (UAVs), will need reevaluation. The technological inability to see, and to some extent hear, through walls will force constant assessment of risk to reconnaissance personnel: Is the enhancement in situational awareness gained by entering a structure worth the considerable increase in risk to those having to make that entry? Specific technological needs will become increasingly available as the Marine Corps becomes more familiar with the demands of urban operations. Adapting equipment largely designed for other environments will tend to be the rule in the mean time.

Urban operations are manpower-intensive. Their character also makes them casualty-intensive. Much can be done to prepare Marine reconnaissance better for such operations in the roughly half a decade that is of interest for this report. Few of those improvements will permit a significant reduction in the raw numbers of reconnaissance Marines and other Marines committed to city interiors, however. The risk of large numbers of wounded and killed will remain until means are found to perform remotely the tasks that only these men can currently accomplish. Developing urban reconnaissance TTP will have immediate and significant benefits for the force. Linking that initial effort to a more long-range vision that incorporates new technologies allowing fewer Marines to be put in harm's way will magnify the influence of those initiatives.

ACKNOWLEDGMENTS

The authors first thank the sponsors of this project, BGen Bill Catto and the men and women of the U.S. Marine Corps Warfighting Laboratory (MCWL), for the opportunity to serve the Marine Corps and the nation at large in undertaking this valuable work. We additionally are most grateful to the many Marines on active duty and retired who granted interviews in support of our research and analysis. Their names appear in the bibliography. Several provided especially notable service in this regard through their efforts to give their time or in helping to coordinate interview sessions. To these gentlemen, an extra thanks: LtCol John Allison, USMC (Ret.), Major Scott Campbell, LtGen Ernest C. Cheatham, USMC (Ret.), LtGen George Ron Christmas, USMC (Ret.), Capt Daniel P. Hinton, SSgt Patrick A. Jirka, Gunnery Sergeant Richard P. "K" Kerkering, Col Michael J. Paulovich, and Col Thomas B. Sward. BG Gideon Avidor, Israeli Defense Forces (Ret.), Major Douglas Chalmers, British Army, and Lt Col Robert Evans, British Army, were also kind enough to grant interviews under exceptional circumstances.

Many of the above also provided informal reviews of the final document in addition to the formal reviews by General Christmas and John Gordon. These efforts are all most appreciated by the authors, as are the exceptional comments made by Frank G. Hoffman of the MCWL.

Dan Sheehan edited the document with skill and care, and Jeri Jackson stepped up to format this document when others were indisposed, and their help is most appreciated, as is that of the ever-reliable woman in the clutch, Terri Perkins.

GLOSSARY

AC-130	A propeller-driven military gunship aircraft
AFB	Air Force base
AFM	Army Field Manual (British)
AH-1	Cobra helicopter gunship
AK-47	Automatic rifle primarily manufactured in former Warsaw Pact nations
ALE	Automatic Link Establishment
AO	Area of operations
ARVN	Army of the Republic of Vietnam
AT-4	A portable antitank grenade launcher
ATV	All-terrain vehicle
BG	Brigadier General (U.S. Army abbreviation)
BGen	Brigadier General (USMC abbreviation)
BLT	Battalion landing team
BMP	Tracked infantry fighting vehicle manufactured in former Warsaw Pact nations
BTR	Wheeled personnel carrier manufactured in the former Warsaw Pact nations
CAS	Close air support
CASCOM	Combined-Arms Support Command
CCIR	Commander's Critical Information Requirements

CH-53	Model of USMC transport helicopter
CINC	Commander in chief
COTS	Commercial off-the-shelf
DTAMS	Digital Terrain and Mapping System
DZ	Drop zone
E&E	Escape and evasion
FAST	Fleet Antiterrorism Security Team
FFP	Final Firing Position
FIRREP	Frequency Interference Report
FM	Field manual
G2/G3	Staff sections in a command led by a general officer. G2 personnel are responsible for intelligence matters, G3 for planning, oversight of operations, and training.
GPS	Global Positioning System
H-250	A tactical radio
HUMINT	Human intelligence
IBCT	Interim Brigade Combat Team
IDF	Israeli Defense Force
JTF	Joint Task Force
LAV	Light Armored Vehicle
LAW	Light Antitank Weapon, a portable and disposable antitank grenade launcher
LD	Line of departure
LOS	Line-of-sight
LP	Listening post
LTC	Lieutenant Colonel (U.S. Army abbreviation)
LtCol	Lieutenant Colonel (USMC abbreviation)
LtGen	Lieutenant General (USMC abbreviation)
LZ	Landing zone

M113	Tracked personnel carrier manufactured primarily in the United States
M16	Automatic rifle manufactured primarily in the United States
M203	Combination automatic rifle-grenade launcher system
M240	A medium machine gun
M4	Automatic carbine manufactured by the United States
MACV	Military Assistance Command, Vietnam
MAGTF	Marine Air-Ground Task Force
MCRP	Marine Corps Reference Publication
MCWL	Marine Corps Warfighting Laboratory
MCWP	Marine Corps Warfighting Publication
MEB	Marine Expeditionary Brigade
MEF	Marine Expeditionary Force
MEU	Marine Expeditionary Unit
MEU(SOC)	Marine Expeditionary Unit, Special Operations Capable
Mk. 19	An automatic grenade launcher
NAI	Named area of interest
NATO	North Atlantic Treaty Organization
NBC	Nuclear, Biological, and Chemical (weapon)
NCO	Noncommissioned officer
NDRI	National Defense Research Institute
NGO	Nongovernmental organization
OCSW	Objective Crew-Served Weapon
OICW	Objective Individual Combat Weapon
OP	Observation post
OPFOR	Opposing force
PSID	Personnel Seismic Intrusion Device

PVO	Private volunteer organization
R&S	Reconnaissance and surveillance
RAILREP	Rail Reconnaissance Report
ROE	Rules of engagement
RPG	Rocket-propelled grenade (primarily used to refer to the launching system)
RSTA	Reconnaissance, Surveillance, and Target Acquisition (squad)
S2X	Position in the IBCT intelligence staff section. The S2X is "responsible for the planning, tracking, and execution of all HUMINT-gathering operations throughout the brigade's area of interest."
SALUTE	A report format: Size, Activity, Location, Unit, Time, and Equipment
SEAL	Sea, Air, and Land. SEALs are a U.S. Navy special operations asset.
SINCGARS	Single-Channel Ground/Air Radio System
SITREP	Situation report
SOP	Standard operating procedure
SOTG	Special Operations Training Group
SPOTREP	Spot Report (or enemy siting report)
SSCC	Scout Sniper Control Center
STA	Surveillance and Target Acquisition
TAI	Target Area of Interest
TOC	Tactical Operations Center
TPT	Training practice tracer
TTP	Tactics, techniques, and procedures
UAV	Unmanned Aerial Vehicle
USMC	U.S. Marine Corps
VC	Vietcong

Chapter One

INTRODUCTION

> Like civilian personnel, civilian buildings and towns normally have a protected status—for example, they are not legitimate targets. Buildings and towns lose their protected status if the appropriate authorities determine that the enemy is using them for military purposes. If doubt exists as to whether a town or building is defended, that doubt should be settled by reconnaissance—not by fire.
>
> British Army *Urban Operations* Field Manual

> Urban reconnaissance is an open book. . . . If you haven't gone out and performed urban reconnaissance . . . you need to go out and perform a series of experiments to determine what is needed.
>
> LtGen G. R. Christmas, USMC (Ret.)

The column moved rapidly down the city avenue, tank battalion leading and two mounted infantry units following. The infantry were unusually equipped, the first of the two units riding in captured BTRs,[1] the next in half-tracks and trucks. It was many days into the war, and some innovative requisitioning had been called for given the distances covered, number of vehicles lost, and ad hoc nature of some units.

The urban area had little tactical value here in the closing days of the conflict, but its strategic value was considerable. The Egyptians' lines of supply ran from its buildings and through its streets. Deny-

[1]Wheeled personnel carriers from the former Warsaw Pact.

ing them the built-up area meant that an entire corps would be stranded in the desert without means of replenishment (Zaken, 2000). Its loss would force their negotiators into the embarrassing situation of having to surrender more at the bargaining table or watching thousands of their soldiers wither under the glare of the desert sun and the international media's spotlight.

The responsible division commander asked his immediate superior whether he should attack into the urban area. He was directed to do so "provided it does not become a Stalingrad situation" (Adan, 1980, p. 409). In retrospect, the Israeli attack on the objective, Suez City, would be far more akin to that infamous World War II battle than any would have imagined. In a war known as yet another stunning Israeli victory, the Battle for Suez "proved to be a very grave error indeed" (Herzog, 1985, p. 282).

The mission to lead the attack fell to Lt. Col. Nachum Zaken's armored battalion. The lack of a reconnaissance effort was thought in retrospect to be a, perhaps the, crucial element in the Israeli defeat. The battalion commander himself would observe that "when you are talking about [operations in] cities . . . you must study every street, every situation, every government building. . . . If you study the city, you can maneuver. If not, it is a matter of luck" (Zaken, 2000).[2] To have gotten a good study of Suez before the operation would have taken five to six hours in Zaken's estimate, still leaving time to attack before the implied suspense of seven the next morning when a cease-fire was to take effect.

The battalion commander had only a 1:50,000-scale map of the city. No air photography was available. Colonel Zaken therefore lacked information on the width of roads, size of buildings, and other crucial details. He asked if he had artillery and air support. Much of the city's civilian population had departed in the years before during the exchanges of artillery fire, commando raids, and air strikes that became known as the War of Attrition.[3] Nevertheless, political sensitivity caused military leaders to restrict the use of supporting air

---

[2]The interview with General Zaken is the primary source for this description of the battle for Suez City.

[3]For a concise synopsis of the War of Attrition, see Herzog (1975, pp. 7–12).

and ground fires.[4] No significant resistance was expected. The adversary was known to have forces in the city, but they were thought to be scattered, lacking in cohesion, challenged in leadership, and suffering the same collapse of morale as had much of the already defeated army. "But," Colonel Zaken recalled, "there was a mistake." The Egyptians had sent a skilled commander into the city with the mission to defend it. "I don't think they made real preparations. They didn't have the time . . . but it was enough" (Zaken, 2000).

The Egyptian leader had arrayed many of his defenders along the avenue that Zaken's force would use to conduct its attack. Concrete walls 80 centimeters high lined both the sides of that road. If a vehicle was hit and rendered immobile, these walls meant it was very likely that none behind could pass until it was moved. The tankers and their trailing infantry would find that it took several efforts and five to ten minutes to breach the walls and bypass immobile vehicles, the men forcing the breach suffering incoming enemy fire for the duration of the frantic efforts. A lack of appropriate maps, overhead imagery, and ground reconnaissance denied the attackers information regarding the foe's dispositions, conditions along the attack route, and other intelligence that would have had a fundamental influence on the planning and execution of the mission.

The attack axis led from the north of the urban area to its central area, Port Ibrahim. The armored battalion moved out between 0830 and 0900, October 23, 1973, with roughly 40 vehicles. Between 40 and 80 meters separated each member of the long line. The column was roughly two kilometers long by their battalion commander's estimate. The tankers' mission was to seize the main street of Suez as far as the port, after which the two trailing infantry battalions would clear the remainder of the built-up area. Looking back, given the size of the city and the level of resistance, Colonel Zaken concluded that "it was impossible" (Zaken, 2000).

There was no resistance as the column passed the buildings on the outskirts of Suez, no sign of the enemy. Such was not the case for long. Small-arms fire began to strike Israeli tanks and their armored

---

[4]The military situation was further complicated by the fact that it was the third ceasefire that was pending, two earlier having collapsed (Zaken, 2000; Adan, 1980, p. 410).

infantry companion vehicles some 500 meters from the city's outer edge as the density of structures increased. The Israelis continued to advance; Colonel Zaken's the third vehicle in line, those of a company commander and the company second in command immediately in front of him. Heavy fire from virtually all directions suddenly impacted the column approximately a kilometer past the first building. Zaken, at this time roughly 100 meters behind the lead vehicle, received casualty reports in rapid succession. Within minutes virtually every company, platoon, and tank commander had been killed or wounded. (All had been traveling with their upper bodies out of their tank hatches.) The tank immediately to the front of Zaken's erupted as it suffered a catastrophic kill, the tank exploding, its turret blown off its chassis. The battalion commander's vehicle rammed the stricken vehicle from behind, throwing Zaken out of his hatch and onto the street. Aware of the target a stationary tank offered, he frantically signaled his driver to keep moving and climbed aboard as the vehicle managed to force its way past the hulk to its front.

Survivors came on the battalion command net, crying in the clear that their comrades lay dead or wounded in great numbers. Some units had suffered as many as 10 killed in a single vehicle. Colonel Zaken had to make a decision: Should he continue the attack toward Port Ibrahim or pull back along the way his force had come? The latter would be difficult because of the number of vehicles destroyed or immobilized that blocked the road. Yet to continue with sporadic communications, his chain of command in tatters, and no idea of what lay ahead was to risk further casualties and disintegration of the force into many mutually unsupporting fragments. The battalion commander decided to forge ahead, making his intentions known with hand and arm signals for those without radio communications.

The lead elements of the armor battalion reached their destination by early afternoon. The battalion headquarters set up in the city square some four kilometers into the built-up area. Remaining armored vehicles and trailing infantry units were spread out on both sides of the road for nearly the entirety of that distance. Maps were insufficient to support calls for fire. The height of surrounding buildings prevented the sight of artillery spotting rounds. Israeli artillery at one point hit Zaken's armored force as it sat on its objective. Egyptian units continued their attacks; complete destruction threatened the three Israeli battalions. There were no surviving medics and no medical supplies beyond those in vehicle aid boxes (which many soldiers did not know how to use effectively). Few men were unwounded. Some tanks had completely exhausted their ammunition. Many knocked out earlier were virtually full, but cross loading was impossible with the continuing incoming fire. Surviving tanks with rounds remaining defended each other by engaging targets on the side of the road opposite themselves, thus taking advantage of the relative standoff distance to achieve greater effect from the elevation of their gun tubes. As most buildings were from four to eight stories high, tanks immediately next to a structure could not raise their main gun barrels sufficiently to engage targets on the upper floors and roofs.

There was still little understanding of the enemy's strength or capabilities. The Egyptian soldiers employed hand grenades, rocket-propelled grenades (RPGs), and machine guns to continued telling effect. Any vehicle turning down side streets immediately lost radio contact. Without help from comrades, it would fall victim to attacks from all sides and above.

Again using hand and arm signals, Colonel Zaken ordered all survivors to assemble in the city square around him. By 1800 the battalion commander's only means of collecting coherent situation reports was to send a runner north and west along the main streets to assess the situation. The unit eventually moved out of Suez to the southwest (Zaken, 2000). The survivors of the two supporting infantry battalions exfiltrated on foot, leaving their vehicles behind and returning to Israeli lines north of Suez under the cover of darkness (Hisdai, 2000). Attacking Suez City without conducting a preliminary reconnaissance had indeed "proved to be a very grave error."

As will be discussed in far greater detail in Chapter Two, a review of current U.S. Marine Corps (USMC) doctrine reveals that it provides very little guidance regarding urban reconnaissance operations. The service is not unique in this regard. Little coverage appears in the nation's Army manuals or in those for foreign English-speaking militaries. Interestingly, the very limited doctrine that does exist is written for the U.S. Army's Interim Brigade Combat Team (IBCT) and is in draft form as of this writing.

Generic reconnaissance doctrine, that applying regardless of the terrain, is, fortunately, far better established. It provides a solid base for the future development of the Marine Corps urban-specific guidance. Yet while the foundation is solid, much of what has yet to be built will differ from what is appropriate for reconnaissance operations in rural environments. Villages, towns, and cities; military installations and training facilities; and stand-alone buildings and underground complexes all present very different, perhaps even unique, challenges for the reconnaissance Marine. The discussion and analysis that follow address these formidable tasks with the objective of supporting the adaptation so vital to operational success.

Improvements are possible within the scope of evolution. No general call for revolutionary change has arisen simply because the environment is an urban one. There is therefore no need to contemplate a sea change in operational doctrine and in the acquisition, training, and organizational structures driven, or at least influenced, by that doctrine. That is not to say that improving USMC preparedness to conduct urban reconnaissance operations will not involve difficult decisions. Several "sacred cows" require revalidation if they are to be retained.

The primary focus of this report is USMC tactical urban ground reconnaissance conducted during combat operations. Aviation operations receive attention only as they directly affect ground reconnaissance undertakings. "Tactical" reconnaissance, for the purposes of this report, is that with direct application to tactical operations. The echelons receiving the bulk of consideration are Surveillance and Target Acquisition (STA) teams, units in the divisional reconnaissance battalion, and force reconnaissance assets. Emphasis is on near-term improvements—those attainable within the next half-decade and influenced by technologies either available or very nearly so. Technological enhancements are not ignored, but primacy is given to doctrinal, leadership, organizational, and training issues.

"Reconnaissance" as used herein is

> a mission undertaken to obtain, by visual observation or other detection methods, information about the activities and resources of an enemy or potential enemy, or to secure data concerning the

> meteorological, hydrographic, or geographic characteristics of a particular area. (JP1-02, 2002, p. 365.)

Further, Marine Corps doctrine recognizes four basic types of reconnaissance: *route, area, zone,* and *force-oriented*:[5]

> Route reconnaissance is a directed effort to obtain detailed information of a specified route and all terrain which the enemy could influence movement along that route. . . . [It] is focused along a specific line of communication, such as a road, railway, or waterway, to provide new or updated information on route conditions and activities along the route.
>
> An area reconnaissance is a directed effort to obtain detailed information concerning the terrain or enemy activity within a prescribed area, such as town, ridge line, woods, or other features critical to operations. An area reconnaissance can be made of a single point, such as a bridge or installation.
>
> A zone reconnaissance is a directed effort to obtain detailed information concerning all routes, obstacles (to include chemical or radiological contamination), terrain, and enemy forces within a zone defined by boundaries. A zone reconnaissance normally is assigned when the enemy situation is vague or when information concerning cross-country trafficability is desired.
>
> A force-oriented reconnaissance is focused not on a geographic area but on a specific fighting organization, wherever it may be or go.

Reconnaissance and surveillance are separate entities, though surveillance activities can be and often are a part of reconnaissance operations. It is notable that the definition of surveillance ("the systematic observation of aerospace, surface or subsurface areas, places, persons, or things, by visual, aural, electronic, photographic, or other means") includes a demographic component ("persons") whereas that for reconnaissance does not.[6] Further, reconnaissance is by definition exclusively related to "the activities and resources of an enemy or potential enemy." The definition should be revised to

---

[5]Definitions are from MCWP 2-15.3, 2002, pp. 1-1–1-2.

[6]The definition for "surveillance" is from JP1-02, 2002, p. 422.

encompass both other-than-combat scenarios and noncombatant considerations. It is noteworthy that Marine reconnaissance units have repeatedly demonstrated their value during stability and support missions in which combat and enemies per se did not exist. Nonetheless, mission requirements mandated collection of information regarding parties with interests that might conflict with or be complementary with those of the United States and its coalition partners.

A final extract from basic Marine ground reconnaissance doctrine serves to emphasize the importance of much of the discussion that follows. Being aware of the "Fundamentals of Ground Reconnaissance" as one considers the analysis presented helps readers to understand how reconnaissance in support of urban operations both demonstrates these basic truths and is at times in tension with other mission demands. These fundamentals are as follows:[7]

- Ground reconnaissance supports the commander's intent and his priority intelligence requirements.
- Ground reconnaissance generally provides highly reliable intelligence information.
- Reconnaissance assets are best employed early to support situation development and friendly course of action development and selection.
- Reconnaissance assets are best employed in general support.
- Reconnaissance requires adequate time for detailed planning and preparation.
- Reconnaissance requires adequate time for execution.
- Reconnaissance must be integrated into the overall intelligence operations plan.
- Effective reconnaissance integrates reconnaissance and intelligence-collection planning.

---

[7]The fundamentals of ground reconnaissance can be found in MCWP 2-15.3, 2002, pp. 1-7–1-12.

- Reconnaissance forces should orient on the enemy to gain and maintain contact.
- The best ground reconnaissance asset should be employed for each specific task.
- Reconnaissance relies on stealth, maneuver, and timely and accurate intelligence reporting.
- An evolving tactical situation requires flexible reporting to the supported command.

This analysis seeks to provide better understanding and improve USMC urban ground combat reconnaissance by first identifying relevant areas in which enhancement is called for and, second, by considering how to achieve those improvements. The latter step is organized into four thematic areas that assist in identifying the character of both the challenges confronting the USMC and potential means of overcoming those challenges. These four thematic areas are:

- The urban environment demands almost constant creative adaptation. Its inherent character, compression of space, and related proximity of participating parties necessitates rapid adjustments in reaction to adversary behavior or to influence that behavior to favor friendly force objectives.
- Tactical ground reconnaissance is a system of systems within a system.
- Urban operations impose extraordinary leadership, training, task organization, and management demands.
- The urban environment makes special demands on equipment and technology. Testing in rural environments does not constitute testing for urban operations.

Chapter Two considers current USMC ground urban reconnaissance deficiencies. These shortfalls appear under one of four headings: doctrine; training; organizational structure, manning, and personnel management; and materiel. Specific shortfalls and observations are the products of a review of pertinent literature, interviews with both retired and active-duty Marines and other persons with relevant experience, and extensive analysis of previous studies regarding military urban operations. The reader may occasionally find this chapter a somewhat bumpy ride. By their very nature, the results cover the entirety of the urban reconnaissance subject area. Some findings are related, making their discussion and analysis easy for author and reader alike. Others are less amenable to reader-friendly presentation, not fitting neatly into any one of the above headings and requiring an assist from a literary shoehorn to make their presentation palatable. The authors have made every effort to retain both clarity and a readable style in these sections, but where style and clarity were in tension, the needs of clarity ruled the day.

To help in the identification of specific urban reconnaissance shortcomings, the major component of each shortfall is presented in boldface type. For ease of reference, the Appendix presents a concise summary of the material shown in bold without its accompanying explanatory material.

The bulk of subsequent analysis takes each of the thematic areas in turn as the basis for considering what is needed as the Marine Corps develops urban ground reconnaissance tactics, techniques, and procedures (TTP) for use during combat operations. The consideration attempts to avoid taxing the reader with a recitation of the obvious. It does not address existent TTP with obvious direct application to urban contingencies. Rather, the objective is to identify and investigate needs unique to or notably influenced by the demands of operating in urban environments.

Every insight into the adversary's dispositions, capabilities, or intent further supports efforts to impose the friendly force's will on the enemy while bettering the chances that Marines survive the mission unscathed. The men in Colonel Zaken's attacking column paid the penalty for a failure to conduct effective urban reconnaissance. Alternatively, good urban reconnaissance and the units that perform it can be the keys to battles won and lives saved. The purpose of what follows is to assist in honing the implement that conducts the mission.

Chapter Two

SHORTFALLS IN USMC URBAN GROUND COMBAT RECONNAISSANCE

DOCTRINE

> To do the clearing of Hue correctly, the Marines first should have isolated the city [which was never completely done by either side as the western side of the urban area was left open]. Second, leaders should have selected the point at which to enter the built-up area. In this case, the selection was predetermined by the mission as we were told to immediately move on MACV headquarters. Third, "make the determination of your clearance technique. The point is, if you have the opportunity . . . to isolate the city and select your entry point, the reconnaissance determines the location of the key terrain that allows isolation. Reconnaissance determines how to isolate the city and [helps to] determine the entry point. Reconnaissance becomes critical in finding the route and determining how to clear the urban area."
>
> LtGen G. R. Christmas, USMC (Ret.)[1]

> Battalion commanders want to move quickly in urban areas, but platoon leaders want to do it slowly.
>
> Brig. Gen. Gideon Avidor, IDF (Ret.)

General

There is much of pertinence to urban operations in existing USMC generic reconnaissance doctrine. However, **formal, written urban**

[1]Paraphrased and quoted (in quotation marks) from Christmas (2001).

combat reconnaissance doctrine—the foundation (at least in theory) for the planning and execution of operations and training, the development of organizational structure, and the basis for equipment procurement—**is essentially nonexistent.** USMC manuals provide virtually no guidance. The NATO reconnaissance publication, *Reconnaissance and Surveillance Support to Joint Operations*, is similarly deficient. British and U.S. Army publications are more forthcoming but still fall far short of comprehensive discussions regarding how to perform reconnaissance properly in densely populated and built-up environments.

Written doctrine normally encompasses field practice based on combat experience and training. These proven, if at times dated, procedures are supplemented by input from those tasked to ensure that the doctrine reflects current requirements given the inevitable evolution of threats, technologies, and capabilities. Sometimes the doctrines of other nations' militaries or procedures employed by related professions provide insights of value.

The concern with regard to urban reconnaissance today, especially involving reconnaissance in support of tactical ground combat operations, is that U.S. armed forces have virtually no experience on which to base written doctrinal guidance. The last operations involving extended ground combat in cities date from the Vietnam War. More modern episodes, such as Mogadishu, Beirut, or Khafji, lack the duration, large-scale participation by units other than special operations forces, or scope of relevant mission requirements to make them bases for general guidance. Lessons learned from other militaries' recent operations in Chechnya, Israel, or elsewhere have some application, but differences in force capabilities, moral perspective, or mission limit their value in developing comprehensive tactical ground combat doctrine. Similarly, insights from fire, police, or other domestic services are generally only relevant to very specific elements of military operations.

The inevitability of future Marine urban operations suggests that it would be wise to address the current doctrinal dearth. Quality training based on sound professional judgment, historical study, and analysis of the contemporary security environment at first seems to be the primary source from which viable urban reconnaissance doctrine can be drawn. Yet notably, both the USMC and the U.S. Army have developed an innovative potential source of doctrinal guidance that has perhaps heretofore not been sufficiently recognized. The Marine Corps Warfighting Laboratory (MCWL) and the several Army Battle Lab facilities conduct activities that involve the testing of tactics, techniques, procedures, and organizational structure in addition to technologies. The results of trials, exercises, and experiments run at these facilities provide a potential supplementary and complementary source of doctrine heretofore rarely available to militaries. Given that their charter encompasses consideration of future operational environments, doctrine developed based on laboratory activities could be especially pertinent to coming deployments of U.S. armed forces.

Deriving doctrine, including TTP, from these organizations must be done with caution, however. The results of these laboratories' efforts should, from a doctrinal sense, be viewed as more illustrative than authoritative because of the virtual impossibility of conducting controlled experiments and achieving repeatable results typical of those completed in academic and scientific laboratory settings. "Generalizable" results in the sense of what would be accepted in accordance with established academic and scientific standards require definition of a hypothesis subject to experiment (e.g., "Procedure X constitutes a new window entry technique that saves time and reduces friendly force casualties"). The hypothesis is then tested repeatedly across an appropriately large sample of windows using multiple control groups, some applying the old procedure and others a new one. The control and experimental groups would ideally be similar in every respect aside from the difference in procedures. A sufficiently large sample of windows and use of different control and experimental groups would be employed to provide the repeatability of results necessary to transcend the experimental results from "exemplar" to "generalizable" for the general population of windows and persons who perform window entries. Otherwise, the results might only be applicable under certain conditions and for certain types of persons—e.g., "The new technique can only be said to be superior for entering bay windows with a force of Marines all of whom are taller than six feet, two inches." Where such qualifications exist, they must be understood and made explicit to avoid drawing inappropriate and unsubstantiated doctrinal inferences or conclusions based on the "test," "trial," or "experiment."

The observations and findings from the just-mentioned military laboratory efforts are nonetheless valuable in identifying what should be considered potential additions to doctrine. The current publication of "X-files" by the MCWL is a step toward disseminating such lessons. Steps should be taken to ensure that they are aggressively distributed to appropriate organizations for consideration and potential validation during training or field operations, after which those passing muster can be incorporated in relevant doctrine.

Much of current USMC reconnaissance doctrine and many traditional reconnaissance procedures apply to missions undertaken in modern urban areas. However, a danger exists that too great a reliance on tried-and-true methods will preclude development of innovative approaches in an environment that many agree puts extraordinary demands on collecting, synthesizing, and disseminating information. It is widely recognized, for example, that **intelligence collection in densely populated areas is more reliant on human intelligence (HUMINT) than is normally the case in other environments. Yet, there is little guidance regarding how Marine commands should integrate this greater reliance on HUMINT into their collection and analyses processes.** Material derived from such human sources—whether local nationals, enemy prisoners of war, or others—should be quickly and effectively incorporated with intelligence otherwise obtained by Marine combat reconnaissance personnel.

Savvy employment of urban target systems analysis and urban intelligence preparation of the battlefield can enhance the value drawn from such HUMINT. Just as these methods serve to identify high-payoff targets or key terrain, they can identify those demographic nodes (e.g., an influential religious or community leader) that can provide HUMINT opportunities with greater mission value than others.[2]

The above should be considered with the following qualification: While historical experiences during Hue and elsewhere have pointed to the unquestionable value of HUMINT during even intense urban combat, there is a greater likelihood that noncombatants will be in

---

[2]Marine Corps Intelligence Activity (1998) provides a number of relevant questions to assist in determining both physical and demographic critical points in built-up areas.

hiding under such circumstances and will therefore be both less well informed and difficult to access than is the case during support or stability missions. It may thus be that HUMINT, while still vital, will play a lesser role during "Block 3" missions in urban areas (Evans, 2001).[3]

There is a need to further investigate the possibility that the complexity of urban areas may impose greater responsibility on Marine teams to provide analysis versus only reporting what is seen (e.g., via a SALUTE report). Limited lines of sight and related fleeting glimpses might make it impossible to observe more than a small fragment of an enemy organization. Marine leaders need to determine whether they prefer urban reconnaissance elements to report what is in their judgment "an enemy dismounted infantry squad reinforced with at least one armored vehicle" rather than what they have direct knowledge of: "two dismounts and sounds of a heavy vehicle." Such a change would put a greater burden on more-junior (in terms of rank, but likely more-experienced) personnel. One intelligence officer noted that during Project Metropolis at George Air Force base (AFB), Calif., Marine "scout sniper teams were identifying individual enemy personnel and fire team–sized (four man) units. With swarms of these personnel swarming over a constricted battlefield, teams were required to focus on the critical information without getting bogged down on the minutia of individual enemy personnel" (Mangan, 2001). The same officer noted that his toughest challenge was "separating the 'wheat from the chaff.' . . . The amount of information passed was staggering!" (Mangan, 2001). The decision is not a straightforward one. What is "chaff" to one consumer of reconnaissance reporting (e.g., a regimental S-2) is "wheat" to another (for example, the infantry squad leader who will encounter the nearby enemy fire team). Doctrine writers would also have to determine whether demanding more analysis from reconnaissance team members could withstand a sudden expansion of the Marine Corps should a major conflict require such an unlikely growth. The training implications of such an expansion are obvious.

---

[3]"Block 3" refers to former Marine Commandant Gen Charles C. Krulak's "three-block war" concept in which a unit operating in a city might be providing support to the population in one block (Block 1), conducting security tasks in the next (Block 2), and have Marines involved in combat in a third (Block 3) (Evans, 2001).

Related to this concern regarding the volume of information was the observation that the density of activities in urban areas is such that traditional reporting procedures may overwhelm reporting and analysis systems. A scout sniper platoon commander noted that "we train Marines to report what they see, and if [they are] trained to report in detail properly they will flood a unit with information" (Ziegler, undated, p. 4). One of his recommendations to redress the issue in part: "Comm[unications] procedures need to be in place for sending FLASH messages when the net is clogged with maneuver units [or reports of lesser importance]" (Ziegler, undated, p. 4). In the future, portable data-processing technologies may reduce the communications and reporting burden by permitting reconnaissance Marines to submit line drawings or imagery depicting multiple adversary positions in a single report sent by burst transmission. While this will help in overcoming problems related to busy frequencies, the burden on those receiving the information will in no way be reduced.

Urban environments may precipitate other fundamental changes. Increased densities of noncombatant and enemy personnel and reduced lines of sight mean that teams will often be in closer proximity to their targets than in open terrain. The chances of teams being detected are therefore greater than on terrain where Marines can detect approaching enemy at a greater distance. Restricted passageways and the possibility that the adversary has (deliberately or incidentally) cut off all viable means of escape will complicate avoiding contact or capture. **It is possible that urban reconnaissance teams will have to be better armed** (e.g., with grenades and Claymore mines) to abet breaking contact or to buy time while vital hardware or intelligence material is destroyed. Additionally, they may be in closer proximity to other friendly forces, forces that in some circumstances could assist a compromised force while in others might complicate escape and evasion due to the danger of fratricide.

The issue of improving Marine reconnaissance armament is related to an ongoing and heated debate within the USMC: the extent to which force reconnaissance assets should predominantly be a "fighting force" (direct action, combat patrol, ready reaction) rather than a "reconnaissance force," with a primary mission of intelligence collection. Weapons training dominates current predeployment training for force reconnaissance company elements. Many in the community believe that reconnaissance skills atrophy during this period, making a unit less capable of performing reconnaissance activities. Further, some Marine Expeditionary Unit (Special Operations Capable) (MEU[SOC]) commanders thereafter view these Marines as a direct action force, one that should be kept at the ready offshore aboard ship. Force reconnaissance attachments, therefore, often have little opportunity to conduct reconnaissance missions during at-sea deployments, or "floats," as Marines refer to them. A concern exists that the direct action mission is nonnegotiable because it has been promised as a Marine capability to combatant commanders during deployments. This need not be perceived as an either-or issue. It was suggested that maintaining but deemphasizing the direct action mission was an appropriate response.

A similar debate has arisen regarding the primary mission for STA team members: Are they "shooters" or intelligence collectors—or both? Several of those interviewed touted the extraordinary value of STA platoon members as intelligence sources while others insisted that such a role detracted from their effectiveness as snipers.

In addition to the need for urban reconnaissance doctrine, there should be a call for its mirror image. **Counterreconnaissance guidance is also lacking** in the USMC literature, as it is in that for NATO, the U.S. Army, and British armed forces.

"A tactical collection plan has to be what ties the various Marine reconnaissance efforts together" (Christmas, 2001). Though urban areas make special demands on Marine reconnaissance elements, as they do on virtually any Marine organization, it should be remembered that this fundamental truth already incorporated in doctrine will remain the bedrock on which to build an urban reconnaissance doctrine.

Specific Observations

There is a need to delineate STA, division reconnaissance, and force reconnaissance responsibilities relative to each other and to provide guidance with regard to their positioning that accounts for lines of sight, supporting fires, and communications limitations in the urban environment. As the quotations below note, traditional concepts regarding divisions of responsibilities articulated in terms of supporting fires or distances are unlikely to be applicable in many urban contingencies:

> By doctrine, [force reconnaissance] should be deep, and we should stay that way. When a unit is in a city we ought to be looking beyond it.
>
> It depends on the level of combat. If it's full combat like in Seoul it is a different situation than if it is . . . involving lesser combat. . . . It depends on what fires you have supporting you. A STA team is at most five kilometers out from its battalion, within 81-mm mortar range [5,720 meters]. Division recon is generally within the artillery fan [ground fires' area of influence]. Force recon could be sent out up to 500 miles.
>
> There is some overlap and it depends on the mission. All could be conceivably operating within a 10-kilometer radius.[4]

Mission, terrain, and available supporting fires will influence reconnaissance element assignments. However, "deep" in open terrain has far different connotations than when in built-up areas. Snipers in 1968 Hue were only a few buildings away from the remainder of their battalion elements. An operation may not involve the deployment of large numbers of friendly forces outside of a built-up area, meaning force reconnaissance teams may be deployed within or in close proximity to other Marine organizations. On the other hand, the generally desirable objective of isolating an urban area (or a portion thereof) will require locating reconnaissance assets so that they can detect and target enemy elements attempting to gain access to the proscribed area. This mission (likely assigned to division or force reconnaissance units) could mean that reconnaissance Marines are outside, on the edge of, or within the urban area of concern.

In addition to this dearth of guidance regarding how to adapt reconnaissance responsibilities, a similar lack of guidance exists on how to coordinate organic and external (in particular clandestine or "black") intelligence-collection assets. The same densities that shorten engagement ranges and distances between forces will at times cause a compression of the distances between Marine organic and non-

[4]All three quotes are from 1st Force Reconnaissance Company interviews.

organic reconnaissance organizations. Standard means of protecting the external elements may be less applicable in urban scenarios. Retired LtGen Christmas, a veteran of intense combat during fighting in 1968 Hue, concluded that "there is nothing worse than having a black unit in your area and you have a lot of restricted fire or no fire areas. . . . You have to know where those teams are or you're going to kill them." He went on to state that there must be greater openness with regard to exchanging information between organic and non-organic reconnaissance elements when operating in urban areas (Christmas, 2001).

While reconnaissance organizations (e.g., STA teams, division reconnaissance battalions, or force reconnaissance companies) will probably continue to provide much of the tactically relevant urban combat intelligence, any unit can provide information of value. This has long been recognized for combat arms organizations, but others that can also provide valuable conduits for HUMINT are too often overlooked. For example, **prebriefings and immediate debriefings of civil affairs and medical personnel working with noncombatants should be incorporated into collection efforts, whether during Block 1, 2, or 3 missions.**

Urban reconnaissance doctrine and training need to better identify requirements of other Marine units they are likely to support. Reconnaissance organizations assisting in determining urban positions for air defense or artillery systems, for example, require training and appropriate references regarding how to determine a building's capacity to withstand a system's weight and the shock of its discharges. Other factors are less obvious. Positioning artillery systems in enclosed open areas (e.g., surrounded by walls but without overhead cover) can cause fatal concussion injuries among gun crews, thus making seemingly attractive concealed locations deadly.

There is a lack of guidance regarding mission-relevant relationships between critical components of the civilian infrastructure. The effects of combat actions in open areas are generally straightforward. For example, the destruction of a village deprives its residents of their shelter. Actions in urban areas have effects that may be far less obvious. Neutralizing a power source to deny power to enemy in the local vicinity may interrupt its supply to friendly occupied territory or result in closure of the city's airport, delaying incoming civil and military support. Similarly, curtailing distribution of water from a purification plant can affect civil and military operations many kilometers distant, both within and outside the urban area. Such "knock-on" influences are found with greater frequency and have more immediate effect in metropolitan areas. They can also be far more difficult to foresee than is the case in less complex rural infrastructures. (Concerns regarding infrastructure had relevance for training also. Several of those interviewed felt that training with regard to how urban infrastructure might influence Marine missions is insufficient. Target analysis training provided by agencies outside the Marine Corps was thought to be "pretty good," but it suffered in that it "needs to focus on other than U.S. cities" and is available only once annually.) A need also exists to understand the social infrastructure—e.g., the doctors, nurses, and other personnel who staff a hospital—rather than considering only the inanimate physical components.

Interviews with 1st Force Reconnaissance Company Marines pointed to **a lack of information regarding how to conduct subterranean reconnaissance.** The British Army Field Manual (AFM), Volume 2, *Operations in Specific Environments,* Part 5, *Urban Operations,* specifically covers subterranean operations in Chapter Four, Part II. While more is needed, these four pages are a good start that provides several points of value. The recently published Change 1 to U.S. Army FM 7-92, *The Infantry Reconnaissance Platoon and Squad (Airborne, Air Assault, Light Infantry),* has a section entitled "Urban Reconnaissance Tactics, Techniques and Procedures" that also offers considerable information that will abet initial steps toward Marine development of reconnaissance TTP, to include those encompassing subterranean operations.[5]

The MCWL and personnel serving in reconnaissance units both recognize **the absence of viable guidance regarding the insertion and extraction of reconnaissance elements.** Use of aircraft (**helicopter support operations are another area seen as requiring much more investigation**) may be unfeasible. The demonstrated vulnerability of rotary-wing airframes (e.g., Mogadishu and Grozny) means that a primary reconnaissance mission may be identification of enemy air

[5]See U.S. Army (2001a). The urban reconnaissance section covers pp. 9-15–9-37.

defense assets within a city. Unless sufficient resolution of this threat can be obtained using overhead assets prior to insertion efforts, the risk of using helicopters to insert the same forces needed to secure safe landing zones will be seen as putting the cart before the horse. While specific tactics (false insertions, rooftop landings) may work early in an operation, their viability is likely to quickly diminish as the foe adapts.

Ground insertion techniques have proven viable for allied forces, the use of which has thus far been denied by some Marine Corps leaders. British reconnaissance elements in Kosovo donned coalition member uniforms and teamed with them for both foot and vehicle patrols through future British Army areas of operation. Similar initiatives to accompany coalition forces during vehicle patrols in Mogadishu were disallowed by Marine leaders for reasons that are unclear.[6] The possibility of capitalizing on such opportunities should not be overlooked in the future. Discussion of these and other ground insertion techniques (including use of indigenous vehicles and drivers and "swarm drills" during which large numbers of a mounted patrol move into an area, subsequently departing with one or more patrols or observation or listening post [OP or LP] teams left behind) is thus far lacking in formal USMC doctrine.[7]

Time factors for urban insertions and extractions are unknown and may vary from those in open terrain. One member of the Marine Corps 2nd Reconnaissance Battalion recognized that the command estimate process for a given mission must first involve a decision regarding the feasibility of committing manned reconnaissance assets, noting that "if you need intelligence immediately you don't send recon out. You use something else. Twelve to 24 hours is about realistic for force recon forewarning."[8] First Lieutenant Brian Ziegler

---

[6]From interview of 1st Force Reconnaissance Company personnel by Russell W. Glenn, Camp Pendleton, California, July 12, 2001.

[7]Interview notes from George AFB, Calif. Interviews by Jamison Jo Medby, February 6–7, 2001. Capt M. Ciancarelli, 2nd Reconnaissance Battalion, similarly noted that members of the British Army's Royal Dragoons would make repeated passes by buildings during "routine" patrols to obtain specific information on targets of particular interest. Frequent halts in the vicinity, especially those at night, would provide opportunities to infiltrate reconnaissance teams (Ciancarelli, 2001).

[8]At the time, the division reconnaissance battalion included the 2nd Marine Expeditionary Force reconnaissance capability.

independently arrived at a similar conclusion based on his experiences as Scout Sniper Platoon Commander during the February 2001 MCWL Project Metropolis experiment in Victorville, California. He concluded that "you would need at the very minimum 36 to 48 hours before the trigger pullers cross the LD. . . . The urban environment doubles the need for stealth, which doubles the usual time needed for movement into the AO," or area of operations (Ziegler, 2001, p. 3). Further evaluation is called for as this "rule of thumb" was drawn from but a single series of events conducted at one location.

There is a need for planning and coordinating fire support plans to cover teams during reconnaissance missions and to minimize the number of changes to those plans during missions. Similarly, urban escape and evasion (E&E) plans should be uniform and coordinated. "Wing, BLT [battalion landing team], and other entities all generate their own E&E plans. They ought to all be the same, or at a minimum coordinated."[9]

Marine air support for ground reconnaissance suffers from the same absence of doctrine and training opportunities as do ground elements. Whether manned or unmanned platforms are employed, the standoff capabilities provided by aircraft and the capabilities inherent in employing such assets in support of the overall reconnaissance mission demand their inclusion in doctrinal guidance. Several of the shortfalls addressed during interviews can be mitigated by proper coordination of ground and air reconnaissance assets. Among the benefits of such coordination:

- Confirmation or updating information on maps and overhead photography.
- Identification of rooftop positions suitable for OPs, LPs, sniper positions, or insertion and extraction points.
- Preliminary reconnaissance of selected structures by looking through windows with visual or thermal capabilities.
- Observation of approaches from the flanks or the far side of the AO, either within or outside the urban area (Schenking, 2001, pp. 15–17).

---

[9] 1st Force Reconnaissance Company interviews.

The density of noncombatants, critical infrastructure nodes, and other indigenous civilian elements that may be encountered during urban operations increasingly task military organizations beyond their organic capabilities. Greater cooperation between military assets and those nongovernmental or private enterprises is necessary. Though to a lesser extent than during stability and support missions, **combat reconnaissance elements may still find themselves reconnoitering in support of multinational and nongovernmental organizations (NGOs) and private volunteer organizations (PVOs). No USMC doctrine currently exists that provides guidance with regard to proper execution of or training for these tasks.**

Force reconnaissance representatives envision themselves as responsible for assisting a commander in shaping his battle. **Marine doctrine needs to discuss how reconnaissance assets can best aid leaders in this shaping activity during actions involving villages, towns, or cities.** Depending on the sympathies of the urban population, responsibilities could include efforts to influence indigenous population behavior in addition to more traditional tasks, such as the following:

- Determining how the enemy is supplied.
- Determining how it executes command and control activities.
- Discovering what the reconnaissance teams could do as part of a greater effort to isolate all or part of the urban terrain.

Force and division reconnaissance personnel also expressed concern regarding a lack of tailored intelligence support, for example, an inability to obtain urban maps of an appropriate scale (e.g., 1:10,000 or larger) or timely overhead imagery.

The close proximity of STA teams and other reconnaissance assets to other friendly units during urban operations means that traditional reporting procedures may be inappropriate. A STA team in general support to its battalion will usually report to the Sniper Control Center (platoon commander), which in turn reports to the battalion S2. Information is then processed and disseminated to appropriate users in the line companies. Because of the density of forces and activities in heavily populated and built-up areas, this process may not be timely enough to serve mission requirements.

There may be a need for modifications to doctrine that ensure that reconnaissance personnel have a broader knowledge of operations in their AO, including the location and mission of units there, and that specify reporting procedures to ensure timely passage of information to both traditional nodes and directly to supported units. The latter may require specification of "trigger lines" that determine at which points in time or space a given reconnaissance element's reporting procedures change (Mangan, 2001).

TRAINING

> Gibler's company commanders helicoptered into the firebase for a briefing. The companies themselves would not close until the next morning. "I kept looking at Saigon on the map," remembered Gibler. . . . "I asked the company commanders, 'When's the last time any of you ever did any instruction in your units about Combat in Cities?' They never had, so I said, 'Well, get in the footlockers and get the manuals out—we're going to have classes tonight on Combat in Cities.'"
>
> Keith William Nolan,
> The Battle for Saigon, Tet 1968

General

Training presents a notable challenge to the Marine Corps reconnaissance community. Selected units receive considerable urban-specific preparation (MEU[SOC] in particular or those supporting MCWL urban experimentation). Others, reconnaissance and infantry units included, undertake far less in the way of such training, the actual extent being a function of commander priorities, inadequate facilities, and many other factors. For example, no urban training facility, and none envisioned, provides the geographical volume or density of challenges necessary to train Marine reconnaissance assets adequately. There is considerable concern with the resultant lack of proficiency in officer and NCO ranks alike. Cognizance of their lack of experience in built-up areas is evident in remarks made by both veterans and serving Marines. The sentiment that "none of us are comfortable in urban because we don't train in urban" environments was widespread. **There is an outstanding and immediate need to develop a comprehensive and tiered approach**

to urban reconnaissance training that incorporates classroom instruction, drills, military training facilities, and actual urban areas. Use of assets such as George AFB near Victorville, California, fulfills part of the fourth element of this requirement, but by themselves such temporary solutions are insufficient. Instruction should include "terrain walks" and other uses of domestic and international densely populated, active civilian urban areas of varying size.

The curriculum and standards for urban training should be consistent in reconnaissance schools and across units. Urban training packages (to prepare units for the specific built-up areas in which they will operate during pending deployments) should be tailored to meet local unit mission requirements.

There is a misunderstanding of weapons effectiveness in cities. The contentions that "mortars will be marginally effective at best" and that "artillery and naval guns have too flat a trajectory to support" urban operations were proved false by Marines in both Hue and Beirut (Root, 2000).

Specific Observations

"Controlling fires is difficult for us," noted a member of 1st Force Recon with the agreement of contemporaries. This was especially true given situations involving avoidance of collateral damage or noncombatant casualties. Urban areas complicate communications, laser designation, and location determination. Finding positions from which to laser designate while at a safe distance and still being in a location that provides an acceptable "cone" for aircraft or artillery engagement requires intimate knowledge of supporting systems, urban geometry, and ways to compensate for the challenges inherent in the latter. Training ground reconnaissance assets is only part of the solution. "Air Force and Navy aircraft fly too high and too fast. Marine air will do what's needed. They know the survivability of their aircraft depends on us and what we're pointing out."[10]

---

[10] 1st Force Recon interviews by Russell W. Glenn, Camp Pendleton, Calif., July 12, 2001.

"Teams are not properly educated with regard to ROE [rules of engagement]. They need to know enough to make the right decision. . . . The quality of ROE guidance is extremely variable [from] mission to mission, from excellent to virtually nil." Further, the ROE have to be robust enough to account for sudden changes in mission. "One minute you're feeding them [the indigenous population], the next you're getting attacked. You can go from Block 1 to Block 3 very quickly."[11]

Cultural awareness/cultural intelligence training for urban reconnaissance personnel was identified as an area requiring significant attention. The MEU(SOC) on-ship preparation sessions were considered too superficial to meet reconnaissance team requirements. This deficiency encompasses knowledge as basic as better instruction in simple language phrases to more sophisticated insights that could provide the basis for immediate decisions of tactical importance. Those specifically cited as examples of the second include the times and dates of periodic events (e.g., prayer sessions, market days), habitual civilian diets, and the expected hours of stores opening and closing. One 1st Force Recon Marine noted that "we need to have a minimal understanding or we waste the first three days just obtaining a basic understanding." Those who recently participated in Project Lincolnia war games share his concerns:

> Cultural Intelligence is very important in urban operations. Thus, relevant local embassy political and tactical information has to be merged with the JTF [joint task force] commander's military intelligence. Also, non-traditional sources of cultural information, such as relief workers, reporters, missionaries, and businessmen, need to be better exploited. ("Lincolnia," 2001, p. 4.)

It should be further noted that cultural awareness is as important to intelligence analysts as to Marines at the "sharp end." The more detailed and comprehensive the cultural understanding, the better analysts can properly interpret the actions of the indigenous population and adversary.

Not surprisingly, several of the areas cited as doctrinal deficiencies were also noted by those interviewed as areas in which reconnais-

---

[11]Ibid.

sance units require better training. These include techniques for inserting and extracting teams and E&E procedures. **Other techniques thought to be of value but insufficiently covered in training are as follows:**

- **Quiet and undetectable urban entry methods (e.g., picking locks and window latches, overcoming computer security systems).**[12]
- **Gaining entry into and "hot wiring" vehicles for use when keys are unavailable.**
- **Better procedures for detecting, neutralizing, and installing booby traps (Root, 2000).**[13]

The Special Operations Training Group (SOTG) was thought to provide the best urban training available in the Marine Corps. However, at least one experienced source considered the instruction to be of "limited application. It is based on a mid-1980s tactical model for support of limited objective raids."[14]

The lack of effective urban training involving units of greater than platoon size was considered a deficiency in USMC readiness.[15]

Though communications, laser designating, photography, and vision enhancement hardware have been improved in recent months, the lack of training that would permit testing these assets in urban environments leaves team members unsure of how built-up areas will influence technological performance during missions.[16]

Urban environments impose special medical concerns for reconnaissance elements. Some require few adjustments other than modifying the contents of personal or corpsman aid packets (e.g., more bandages to account for the increased likelihood of cuts; the

---

[12]1st Reconnaissance Battalion interviews.

[13]1st Force Recon interviews.

[14]From Russell W. Glenn interviews. Source will remain anonymous.

[15]1st Force Recon interviews.

[16]Ibid.

related need for Marines other than corpsmen to sew up such cuts requires more extensive adaptation of doctrine and training). Others require study and special training. Removing casualties from confined spaces and from under collapsed structural material was cited as one such category.[17]

Though unlikely to be the case during high-intensity combat operations, **Marine reconnaissance training is currently too reliant on host nation support,** in the view of several of those interviewed. The result is that Marines are unsure of what will serve as effective means of moving or maneuvering in international urban areas.

ORGANIZATIONAL STRUCTURE, MANNING, AND PERSONNEL MANAGEMENT

> Enhance capabilities to operate in urban and austere environments across the spectrum of conflict while simultaneously further reducing our dependence on existing infrastructure.
>
> Gen J. I. Jones
> Marine Corps Strategy 21

General

Members of the reconnaissance community are unsure of what the optimum standard size should be for reconnaissance teams operating in urban areas. The trade-off between detection avoidance and sufficient combat power should a team suffer compromise came up repeatedly during Project Metropolis discussions and interviews conducted in support of this research. In the absence of operational experience, comprehensive training, or specific testing, no one was comfortable with making a conclusive statement regarding force structure or task organization. Some of the views expressed in this regard reveal relevant concerns:[18]

- "I wouldn't want to [perform urban reconnaissance] with our current six-man team, which is often five men at current

---

[17]1st Reconnaissance Battalion interviews.

[18]Comments are from 1st Force Recon interviews unless otherwise noted.

strength, or even four. I'd want to take twelve men, two teams, or at least eight men [to provide more] combat fire."

- "The team needs flank security and enough in the way of manpower to carry a guy if he is wounded. After a casualty, you've been compromised, so you are no longer recon; you're a combat patrol."
- "I don't think we need to alter our force structure, [but] none of us are comfortable in urban because we don't train in urban."
- "Until we physically go do it, we don't know what we need."
- "In general, units were more effective as two-man teams for the short-duration missions they were assigned [during Project Metropolis at George AFB]. They could cover more NAIs [named areas of interest], traveled with a decreased signature (two vice four men), and provided increased mutual security. Two teams could cover the buildings and approaches to their adjacent teams better. However, it was recognized that operating as a two-man team accepted a degree of risk. Two-man teams would not have been sustainable if the mission lasted longer than four to six hours and would likely have weighed the Marines down significantly (carrying radios, ammunition, and other required gear)." (Mangan, 2001, p. 2.)
- "We deployed two-man sniper teams in Hue, and they were the most supportable. They often operated independently, away from the battalion and company positions or front lines." (Christmas, 2002.)

"The limited line of sight [LOS] from any one position restricted the amount of area one R&S [reconnaissance and surveillance] team can cover. This requires a saturation of R&S teams to cover a particular area or roving team that moves within a building to cover different NAIs. . . . When we sent out two-man teams, we usually started them off [as] four-man sections for an initial penetration into the AO. They would split into two-man teams after the penetration and would also mutually support each other via bounding movements and for a small reaction force. These two teams could also link up if the operation required an extended time in the AO for rest in a harbor site" (Ziegler, 2001, p. 1–2). (An obvious implication of this observation is that more smaller teams can cover a greater number of NAIs than a lesser quantity of larger teams.)

It should be noted that a reconnaissance element generally seeks to avoid detection and combat (barring assignment of a mission that explicitly calls for same). Its task is to observe, report, perhaps perform targeting, and return undetected. Concerns regarding survivability of such assets are valid, but those born of an expectation of fighting as a primary responsibility may be misguided.

That **it is necessary to "break the wall between the G2 and G3"** was a repeated observation. **Information of value to maneuver units at times never reached the elements most in need of it because intelligence personnel too slowly disseminated key information from reconnaissance reports.** The recommendation applied to more than reconnaissance products, however. It was felt that oversensitivity to classification or handling issues caused valuable material from other sources to be kept from those who could best put it to use (or whose survival depended on the information).

Specific Observations

There is a lack of specific information regarding urban infrastructure in mission areas and local national points of contact that can address specific related mission concerns. Members of 1st Force Recon noted that SEALs have access to databases that provide detailed information on city infrastructures, including city sewerage plans and contact information for key design and maintenance personnel. SEALs both actively input and extract data from these sources. Marine reconnaissance units should have access to such databases. In exchange, Marine reconnaissance elements could act as additional sources of input. (Authors' note: Despite the observation made in the field, in truth these databases are available through the MEF G2 (Christmas, 2002). That those in the force reconnaissance company were unaware of this implies that this availability should be better advertised.)

The echelon to which unmanned aerial vehicles (UAVs) will be allocated and how they will be integrated into reconnaissance and intelligence dissemination systems need to be determined. At least one individual interviewed felt strongly that UAVs should not be organic components of reconnaissance units because the manpower, communications equipment, and logistics tail associated with the systems would too greatly burden them. Increases in manning would have to include those operating the equipment, maintenance personnel, and analysts. Manning analyst positions with qualified personnel would be especially important. As noted by a British officer with extensive reconnaissance experience in Northern Ireland, when it comes to the evaluation of real-time visual imagery, "the money is in the analysts who watch the screen" (Chalmers, 2001). It should be noted, however, that assigning the systems to higher headquarters or supporting organizations may hamper the timely transmission of intelligence gained from UAV missions to users in the field.

Members of the USMC reconnaissance community expressed concern that once the difficult task of undetected insertion had been successfully accomplished, **means to resupply Marines in hides, OPs, or LPs without compromising the position are lacking.** Apart from short-duration missions during which a team can carry its sustenance and other support, this logistical shortfall presents a serious problem. Potential solutions include premission establishment of caches, surreptitious resupply drops by mounted patrols, and subterranean resupply. In the not-too-distant future, robotic resupply will likely be feasible.[19]

MATERIEL

> The map was another problem in itself. It really bothered me. I had never been issued any other map like it during my entire lifetime-long three-month tour in Vietnam. The maps we had used humping through the rice paddies and mountainous jungle terrain of I Corps had always been 1:50,000 terrain maps.
>
> Nick Warr,
> Phase Line Green: The Battle for Hue, 1968

---

[19]1st Force Recon and 1st Reconnaissance Battalion interviews.

General

Development of innovative technologies and improvements to those already fielded is a focal point for both USMC and U.S. Army warfighting and battle labs. Given this focus and the guidance of the Commanding General of the MCWL not to overemphasize technological solutions in this study, RAND efforts to identify areas of current reconnaissance shortfalls deliberately avoided covering ground already considered by previous and ongoing MCWL investigations. Members of the Marine reconnaissance community who were interviewed are aware of MCWL technological initiatives and find value in many of them. **There is, however, a concern that too great a reliance on extant commercial off-the-shelf (COTS), military off-the-shelf, or brass board (in advanced concept, early development, or prototype form) products may fail to fully address identified needs in the interest of cost savings or immediacy of fielding.**

Specific Observations

A need exists for acoustic or motion sensors that assist in detecting targets and potential threats in built-up areas. Crude predecessors of such devices proved helpful to reconnaissance elements in Vietnam, for example, where force reconnaissance units employed Personnel Seismic Intrusion Devices (PSIDs) to detect enemy intrusion, notably during periods when a team was in a harbor site. A PSID system had five primary components: four battery-powered transmitters and one receiver. The transmitters would be placed at appropriate locations to provide warning of an approach. Each transmitter would emit its own coded signal so that the Marine listening on the receiver knew from which of the four sensors a signal was coming. Employed in conjunction with Claymore mines, the PSID enhanced security and team effectiveness. Unfortunately, the devices were not infallible. Thunder, rain, artillery, or animals could provide false detections that led to unnecessary alerts (Norton, 1992, pp. 116, 143). Cities pose even greater challenges for such technologies. The density of enemy and friendly forces, noncombatants (even those seeking to do nothing other than avoid those fighting), and vehicles makes it difficult to place sensors to best give readings of value. Sensors need to be disguised to avoid being compromised and removed or placed elsewhere.

The increased opportunity for compromise, given the density of combatants and noncombatants, means that any such devices must be wireless. Even if placement difficulties are overcome, recent modeling of acoustic sensors at RAND reflects that fewer than 25 percent of passing vehicles in heavily trafficked areas are acquired, much less properly identified (Matsumura et al., 2000, pp. 7, 39–44). These findings were for sensors placed in less densely occupied terrain than is the case in urban areas, terrain with fewer challenges related to hard surfaces reflecting noise or vibrations and with less traffic density than is likely to be found in a village, town, or city. Sensors may eventually be of considerable assistance to U.S. Marine urban reconnaissance teams, but it would be unwise to expect too much from these assets in the immediate future. For the next several years, these devices will be of questionable value at best to reconnaissance Marines given their inconsistent cuing data and the potential for compromise that exists when placing such devices in the vicinity of a ground reconnaissance team.

In addition to sensors, **other wireless listening devices, including those that can amplify sounds over considerable distances or distinguish sounds through walls, would permit standoff collection of intelligence.**[20] Simple and economical amplification devices have been readily available on the commercial market for several years.

Design standards for equipment should consider the special demands urban environments put on end items. While an overgeneralization, the observation that moisture (environmental and human perspiration) is a primary cause of failures in much of nonurban terrain, whereas shocks, crushing, other forms of breakage, and dust present the primary challenges in built-up areas, contains some truth. Similarly, urban undertakings place unusual requirements on weapons, munitions, and other equipment. Rounds with limited penetration properties, grenades that detonate on impact rather than after a time delay, nonlethal capabilities, and suppression of weapons sounds for use in near-silent kills of dogs or other targets is a sampling of needs cited by various sources. Many of these capabilities will have application to other-than-urban contin-

---

[20] 1st Reconnaissance Battalion interviews.

gencies as well, but they have been noted as being of particular value during urban combat taskings.

Reconnaissance Marines are pleased that they can often acquire new technologies quickly, but **new equipment purchases are too often not accompanied by the operator and maintenance training necessary to properly employ it, which causes some concern.** Acquisition of equipment, whether through routine channels or COTS, should be integrated into an "acquisition system" that encompasses consideration of how the new item will be integrated into Marine doctrine, training, support requirements, and the employment of other systems.

Reliable communications and Global Positioning System (GPS) signals are areas of notable concern. Traditional force reconnaissance and division reconnaissance missions rely heavily on satellite and high-frequency equipment for communications. The likelihood that these forces will be in closer proximity to parent and supported headquarters during urban missions may mitigate reconnaissance team reliance on over-the-horizon communications equipment. However, the successful use of any line-of-sight system (e.g., Single-Channel Ground/Air Radio System [SINCGARS]) in an urban environment often requires exactly that, unobstructed LOS, to overcome the signal attenuation effects of intervening structures. Recent MCWL experiments have demonstrated that attaining sufficient unobstructed LOS to achieve effective communications can be problematic in urban areas, leading to elaborate networks of radio relays and other workarounds to improve communications reliability. These ad hoc approaches will not always be feasible in a combat environment. Field-expedient and directional antennas, for example, are at times not employed because their detection would compromise user positions. LOS limitations also exist with GPS. They are of notable concern when reconnaissance Marines must traverse underground facilities, such as subways and sewer systems.

Whether solutions to these problems involve common Marine communications systems or specialized equipment appears secondary to ensuring that reconnaissance Marines possess the appropriate systems to perform their most basic function in urban environments: timely reporting of enemy activity. However, an implicit requirement is communications compatibility with nearby U.S. Marines and other forces in the area. Interviews with reconnaissance Marines revealed concerns that their communications systems (such as Automatic Link Establishment [ALE] high-frequency radios) are not compatible with radios presently used by organizations with which they must operate. **"Compatibility even within the Marine Corps is terrible," much less with elements from other services.**[21]

A requirement exists for a stealthier means of monitoring radios. The H-250 radio handset used on Marine radios is considered too loud by some reconnaissance personnel, thus creating a potential source of compromise that threatens teams' security. It was suggested that the listening mode be made substantially quieter.[22] Another option might be to exploit other phenomenology, such as vibrations, to alert reconnaissance Marines that traffic is incoming.[23]

The cumulative bulk of equipment was cited as a concern, one with special implications for urban operations. The need to pass through windows, mouse holes, or other restricted passageways typical of urban terrain led to calls for longer, narrower, "body hugging" means of loading equipment in place of those that protruded beyond the Marine's frame to his right or left.[24] Further, equipment, including boots, needs to be quieter to permit traversing populated areas without being detected.

Several concerns have arisen regarding unmanned aerial vehicles in addition to those already mentioned, most of which are well known to the MCWL. Interviews included calls for

- better system optics,
- the ability to make a visual record of missions,
- solving problems with operating the systems in even moderate wind conditions, and
- revising reporting procedures for disseminating intelligence gathered from flights (e.g., transmission to local units as well as

[21]1st Reconnaissance Battalion interviews

[22]Ibid.

[23]Ibid.

[24]1st Force Recon interviews.

the headquarters or intelligence section assigning aircraft missions).

All of the above have been repeatedly cited and represent but a small number of the considerations that should be reviewed as the appropriate roles for unmanned aircraft are introduced into Marine Corps doctrine.

Miscellaneous Calls for Technological Capabilities

- Portable water purification system.
- Urban Digital Terrain and Mapping System (DTAMS) that allows 360-degree views of selected terrain features for use during planning and rehearsals. Ideally this system could also send three-dimensional terrain representations to a team in the field when necessary.
- A means of accessing existing phone lines and using them for encrypted transmissions.
- Longer-lasting, lighter batteries with no hazardous materials that are capable of working in an airless environment.
- Power transformers (alternatively, equipment that operates using 110-volt, 220-volt, or other power sources).
- A fiber-optic capability to see around corners, under doors, or through windows without exposing the user.
- An effective night photography capability (Campbell, 2001).

USMC URBAN GROUND COMBAT RECONNAISSANCE SHORTFALLS: CONCLUSION

> Interviewer to reconnaissance Marine: "So you're telling me you can't do the job?
>
> Marine's response: "I know it sounds bad, but that's the case."

It has been repeatedly noted that much of current generic USMC reconnaissance doctrine applies to urban operations. It has similarly been brought to the reader's attention that much needs to be done in the four areas of consideration before Marines have guidance, organizations, preparation, and equipment appropriate to the extraordinary demands the urban environment imposes on reconnaissance personnel. The focus of this report is tactical urban ground combat reconnaissance. The lack of recent USMC experience in this realm, and a similar lack of adequately challenging training, leaves the limitations of equipment unknown, questions unanswered, and other problems undiscovered.

An open mind free of predispositions is an essential tool for all seeking to redress these shortcomings. That a definitive break exists between Block 2 urban reconnaissance and reconnaissance conducted during Block 3 missions was a point of major emphasis during the March 13–14, 2001, Urban Reconnaissance Conference in San Diego. The latter were accepted as being inherently harder, with the definition of Block 3 requirements correspondingly more difficult. Historical examples and discussions with those currently serving in Marine reconnaissance units cast some doubt on the uniform applicability of this conclusion. Infiltrating and maintaining the viability of teams during Block 2 missions will be extraordinarily challenging in many circumstances, notably so given the demographic character of reconnaissance units compared with those of the indigenous urban populations into which they are likely to be committed. A Block 3 environment may offer a reduced likelihood of compromise by noncombatants because those individuals will probably be more concerned with their own protection than providing information to combatants. Stalingrad and other examples from the past demonstrate that urban reconnaissance during Block 3 is very difficult and exceptionally dangerous. Nevertheless, it may offer reconnaissance elements more options than are available during Block 2 commitments. Given so little in the way of recent historical experience and applicable training, it falls to all who are seeking solutions to question even the seemingly most obvious conclusions. Further, as is true with reconnaissance activities anywhere at any time, the solutions sought should neither seek to attain nor promise perfection. Neither should leaders believe that demands on their initiative and innovative thinking will be any less during operations because the terrain is dominated by man-made features or a noncombatant population. Major General Baron von Freytag-Loringhoven's observation is no less applicable for those committed to success during tomorrow's urban undertakings than it proved prescient for operations during World War I and World War II after being written in the first decade of the twentieth century:

> If the great generals at Marengo, Ulm, Jena, and Koeniggraetz had waited for the situation to clear up fully, they would have missed the proper moment for action, and military history would be without some of its most brilliant days. (von Freytag-Loringhoven, 1938, p. 79.)

The goal for Marines performing urban reconnaissance is to provide sufficient timely intelligence to establish the conditions for mission accomplishment at the minimum feasible cost in friendly force and noncombatant lives.

Chapter Three

URBAN GROUND COMBAT TACTICS, TECHNIQUES, AND PROCEDURES CONSIDERATIONS

The material in this chapter addresses many of the shortfalls identified in Chapter Two with the objective of establishing a foundation for preparing USMC reconnaissance units for the challenges of urban ground combat in the immediate future. There will be occasional repetition of points discussed in the previous chapter. This allows us to make reference to previously raised issues without having to constantly resort to the phrases, "as noted," "as already mentioned," or others of a similar nature. Chapter Three reviews each of the themes identified on p. 9 in turn.

THE DEMANDS OF THE ENVIRONMENT

Introduction to the First Theme

First Theme. The urban environment demands almost constant creative adaptation. Its inherent character, compression of space, and related proximity of participating parties necessitates rapid adjustments in reaction to adversaries' behavior or to influence that behavior favorably to friendly force objectives.

> Take your time. Stay away from the easy going. Never go the same way twice.
>
> Gunnery Sergeant Charles C. Arndt,
> USMC, Guadalcanal, 1942

> The USMC will not use recon assets for any innovative applications.
>
> Marine during interview with author

The restricted LOS and large numbers of HUMINT sources in urban areas precipitate a more intensive use of manpower than an equivalent grid square on more open terrain. Those myriad human beings so valuable for intelligence collection can also make clandestine operations difficult if not impossible. Insertion of reconnaissance teams and establishment of OPs without detection or preventing compromise while maintaining these capabilities becomes a complex operation in and of itself. Limited LOS means that a team's movement after the initial insertion may be essential to maintain situational awareness beyond the immediate vicinity. The density of noncombatants and enemy assets that can compromise a team during this movement means that shorter moves are desirable. More teams, perhaps many more teams, are needed to cover an operational area than in fields, forests, deserts, or mountains. Ideally, reconnaissance units and intelligence officers would be able to integrate OP fields of view much like a rifle company commander coordinates fields of fire.[1]

The many effects of the urban landscape influence more than reconnaissance units. Whereas artillery support may be sufficient to cover the withdrawal of a reconnaissance team outside of a built-up area, in cities buildings interdict rounds in flight and create large areas of dead space. A combination of artillery, mortars, aircraft, and ground forces may therefore be needed to cover fire support contingencies fundamental to ensuring that a viable escape and evasion plan is in place. Even with these forces in combination, significant coverage gaps will probably still exist. Risk of compromise influences logistical efforts. Either diverse means of resupply must be put in place or the amount of time a team spends on a patrol or manning an OP must be reduced.

The challenges imposed by the urban environment have an immediate and significant effect on signal receipt and transmission. GPS signals may not penetrate buildings or subterranean facilities. Transmissions of voice, text, or photograph traffic are frequently blocked. The routine radio communications on which units rely may fail altogether. Laminated windows can block transmissions. Struc-

---

[1]This interesting analogy was drawn by a participant in the MCWL's March 13–14, 2001, Urban Ground Reconnaissance Conference held in San Diego.

tures might not allow acquisition of the necessary takeoff angle for satellite communications (Miller, 2001; Rutan, 2001). See Table 3.1 for a sampling of various urban sound degradation effects (Edwards, 2001, p. 9). The potential consequences go beyond simply a delay in passing vital information. A Marine participating in the March 2001 MCWL urban experiment at Victorville, California, subsequently observed that "communications failures caused loss of life. Without information coming from the TOC [Tactical Operations Center], scouts and snipers became disoriented . . . and died." The lesson is obvious: As in any environment, reconnaissance personnel must select their positions based on observation, the need to communicate, and access and egress routes. In urban areas, however, locations that provide the essential combination of these characteristics may be far more difficult or impossible to find.

Alternative means of relaying signals is called for. Interestingly, both the Marines' *Reconnaissance Reports Guide* and *Ground Reconnais-*

Table 3.1

Average Signal Loss Measurements Caused by Common Building Materials

Material Type	Loss (dB)	Frequency
All metal	26	815 MHz
Aluminum siding	20.4	815 MHz
Foil insulation	3.9	815 MHz
Concrete block wall	13	1,300 MHz
Loss from one floor	20–30	1,300 MHz
Loss from one floor and one wall	40–50	1,300 MHz
Light textile inventory	3–5	1,300 MHz
Chain-like fenced-in area 20 feet high containing tools, inventory, and people	5–12	1,300 MHz
Metal blanket: 12 square feet	4–7	1,300 MHz
Light machinery: less than 10 square feet	1–4	1,300 MHz
General machinery: 10–20 square feet	5–10	1,300 MHz
Heavy machinery: more than 20 square feet	10–12	1,300 MHz
Ceiling duct	1–8	1,300 MHz
Concrete floor	10	1,300 MHz
0.6 square meter reinforced concrete pillar	12–14	1,300 MHz
Sheetrock (three-eighth inches): 2 sheets	5	57.6 GHz
Dry plywood (three-quarter inches): 1 sheet	1	9.6 GHz
Wet plywood (one-half inches): 1 sheet	19	9.6 GHz
Aluminum (one-eighth inches): 1 sheet	47	9.6 GHz

sance acknowledge the difficulties related to signal transmission, but the doctrine makes no recommendations regarding how to handle or report signal degradation in urban areas specifically. Using separate teams, one to observe and another (or others) to report, is a near-term solution: One team maintains eyes-on an NAI while another mans a position allowing communication with both that OP and the node to which it needs to report. (The increased chance of compromise based on more teams in the field is an obvious drawback of this alternative.) Field-expedient antennas potentially address some of these circumstances, but those in the field have at times found the antenna's signature and resultant possibility of detection by the adversary too great to risk (Ziegler, 2001, p. 3).

Site selection should also account for an environment's electromagnetic characteristics. Tunnels, underpasses, power lines, and large steel structures, such as bridges and metal-framed buildings, can disrupt communications even if LOS with other nodes exists (Edwards, 2001, p. 19). If a Marine chooses to deploy an antenna, he may need to try several directional variations. The seemingly obvious choice—pointing toward the receiver—may not be the most effective course of action. Obstacle amplification or other local characteristics can cause a counterintuitive choice to be the best available (Edwards, 2001, p. 11). During operations in Grozny, the Russians learned to use directional antennas to reflect radio waves off stone or brick walls, thereby propagating signals down a street. When necessary, the Russians placed transmitters and receivers along the routes (Edwards, 2001, p. 21). Given appropriate training, Marines can likewise capitalize on these techniques during future operations.

Reconnaissance units can also take advantage of in-place capabilities. Commercial communications infrastructure may still be operational, providing Marines a landline system that could provide an alternative means (albeit nonsecure) of reporting. (Note that the use of indigenous resources can go beyond communications needs. Capitalizing on available water supplies, fuel, foodstuffs, and the like allows reconnaissance teams to dedicate the weight and space taken by such materials to other needs or potentially extend the duration of their deployments.)

Some Marine units are using portable computers to establish secure "chat rooms" that operate in a manner similar to instant messaging.

This allows all net users to monitor messages in what is effectively real time (Givens, 2001; Lewis, 2001; Edwards, 2001, p. 15).[2]

LOS issues encompass far more than communications issues. Most urban engagements occur at less than 100 meters. A compromised reconnaissance unit therefore has very little time to evade, and the number of routes available for escape may be few. As was noted by a Marine responsible for force protection in Somalia, in urban areas "time and space are compressed. [In other environments, you have warning at greater distances, such as dust on the horizon.] In an urban environment, you don't have that. Things creep up on you without warning. . . . It's much more dangerous" (Allison, 2001). As a result, a contingency plan for reinforcing or extracting a compromised urban reconnaissance team remains a constant need (Allison, 2001). Helicopter, vehicle, and foot movement may all be difficult. As noted, supporting fires may be unable to provide the firepower necessary to break contact. These complicating factors together mean that planning and execution of relief operations takes on significantly greater complexity than would otherwise be the case. Laser employment for designating targets similarly takes on greater difficulties in the midst of buildings. Laser designation requires the designator to have an appropriate angle from his position and the approaching aircraft or munitions to be provided an acceptable angle and duration of flight for acquisition and locking on to the aiming point. The designator and launch platform need to communicate with each other to coordinate their actions. The Marine designating from the ground or air must avoid laser obscuration or its redirection by smoke, dust, or reflective surfaces. All of these favorable conditions need to come together at a time and location where the target is available for engagement. Passing a target from one designator to another may be the only way to destroy a target if it is moving, a task of potentially extraordinary difficulty.

Inherent in the foregoing discussion is a trade-off between the number of reconnaissance tasks that support a commander's concept of operation and the amount of reconnaissance resources available. A single reconnaissance team may be able to observe several cubic

---

[2]Note that there are times when such a universally open system may be undesirable for security, morale, or other reasons. In such cases, a leader could employ alternative chat rooms, whose addresses are disseminated only to a limited number of users.

kilometers on open ground. LOS in very restricted jungle and mountain terrain can also be very limited, but it is likely that viable mobility corridors and avenues of approach are similarly limited, thus making allocation of reconnaissance assets fairly straightforward (though their ability to simultaneously observe air traffic may be restricted). In a city, the LOSs are short and the number of mobility corridors is very high. Tens or hundreds of possible exterior, interior, or subterranean approaches confront a commander. Even the highest point in a city cannot provide observation of the many passages in the lee of structures or within those buildings.

Determining NAIs and other critical points will be a most demanding task. Further, tracking a target will be no less demanding once it is acquired. The doctrinal guidance for coordinating NAI and target areas of interest (TAIs) may require adaptation for urban contingencies. The number of possible alternatives available to a moving enemy force means that designating a TAI and expecting engagement systems to move to and immediately acquire the target may not be reasonable. Doctrine's current two-step process—identify and report on a target at an NAI and engage the target on arrival at a related TAI—might fail to meet operational demands. There is perhaps a call for an intermittent step akin to "move to 'area of target acquisition.'" That acquisition could be a time-consuming and hazardous activity given the concentration of enemy firepower that characterizes much of urban combat.

Commanders and their staffs responsible for collecting and processing reconnaissance teams' incoming intelligence reports will be further challenged by the volume of information those teams provide. The density of activities in built-up areas and the related proximity of friendly, threat, and noncombatant elements mean that the frequency of reports and quantity of information contained in those reports will tend to be greater than is found elsewhere. The volume is magnified yet further given the above-noted potential for commitment of even larger numbers of teams. The resultant burden has been in evidence even in an experimental environment.

Returning to MCWL trials conducted in Victorville, Capt Bryan Mangan, the S2 for the friendly force battalion, noted that "the hardest part . . . was taking the reporting that I received and making decisions based on reports of nothing larger than squad-sized units about [potential enemy] courses of action." In keeping with the observations made above, the same officer further noted that with the assets available in his section, he "had difficulty establishing NAIs that would give me indications of these courses of action" (Mangan, 2001). These difficulties can hurt the quality and timeliness of intelligence products. They impair the ability of maneuver units to engage and destroy targets. Delays in information processing mean that vital information essential to engagement or force protection will arrive too late to be of value. Captain Mangan, after finding his S2 section virtually overwhelmed because of its inability to process all incoming reports, proposed a procedure designed to reduce this time lag. In retrospect, he "determined that a series of 'trigger lines' will need to be established that delineate when reporting goes directly to the battalion [via the Scout Sniper Control Center (SSCC)] and when it goes directly to the supported company commander. When the lead trace of the supported unit (e.g., a rifle company) crossed the trigger line, all scout sniper/scout assets would report directly to that supported commander" (Mangan, 2001).

It may be advisable to refine this process such that reconnaissance team members report select information directly to the units in greatest need, units that may be below company in size. Alternatively, developing appropriate TTP to take advantage of technologies that allow multiple units to monitor significant communications (e.g., common radio frequencies, e-mail access, or instant messaging) could address the need.

Mission-Specific Urban Reconnaissance Adaptation

> When the battalion was able to get ground recon units into the area, the unit did much better and was better able to control the situation.
>
> LTC Fred Gellert, Project Metropolis notes

These many challenges directly affect fundamental USMC urban reconnaissance tactics, techniques, and procedures. LOS issues complicate intelligence collection and both target monitoring and engagement. Shielding, building material, and electromagnetic sources impede communications. The density of enemy combines with the harsh urban terrain to severely tax virtually every support system. These many factors are made yet more difficult by one that simultaneously influences all three levels of war: noncombatants. Civilians flowed out of hiding places to seek the protection of Marines attacking to retake 1968 Hue, draining manpower and offering a way for enemy agents to penetrate friendly lines. Supreme Commander, Allied Forces Europe, Gen. Dwight Eisenhower resisted sending forces into Paris in 1944, fully cognizant that to do so would require supporting the civilian citizenry and thereby drain his ground forces of fuel, food, and other support essential to continued offensive operations. Claims by an international journalist regarding noncombatant suffering in Sarajevo caused a reorientation of military forces and relief efforts during U.S. 1990s operations in the Balkans. And so it is at the lowest tactical levels. Every civilian is a potential source of information for either the friendly force, the enemy, or both. For organizations whose success depends on undetected movement and placement, even a toddler poses an operational threat. Urban reconnaissance would be taxing or difficult even in abandoned urban areas devoid of noncombatants. The presence of civilians increases the level of challenge immeasurably.

Current Marine Corps reconnaissance doctrine has much to offer those undertaking missions in urban areas, but there is call for innovation, initiative, and intelligence in adapting that guidance. There is also a requirement to revise existent doctrine to include urban reconnaissance TTP and concepts that differ from the norm. The following sections consider some of the elements that adaptation and revision might incorporate.

Urban Reconnaissance Infiltration

> Infiltration is the most difficult task.
>
> Brig. Gen. Gadi Shamni,
> Head of IDF Infantry and Paratrooper Doctrine

> I don't think military people can do it.
>
> Col Thomas B. Sward, USMC

> The hardest part is getting in.
>
> Col Michael J. Paulovich, USMC

> The problem is getting into a position . . . without compromise.
>
> Gunnery Sergeant Richard T. Kerkering, USMC

Urban environments influence every facet of reconnaissance operations. This is never more true than with regard to the infiltration of STA teams or elements from divisional reconnaissance battalions and force reconnaissance companies. Many of those representing these units expressed belief that infiltration into an urban area was the most difficult task confronting them. Statements to that effect applied regardless of whether or not stability, support, offensive, or defensive tasks dominated a mission. Successful infiltration might require that the team go in with a smaller and lighter load to reduce the chances of compromise. This directly affects the duration the team can stay on location. Unfortunately, the problems do not end after the completion of an infiltration. Children playing and adults going about their daily business can compromise carefully inserted Marine teams. It has been noted that resupply introduces an additional opportunity for discovery of OPs or other Marine positions. Yet those interviewed believed that once a team reached its destination, the hardest part of the mission was behind them. Getting there was more than half the battle.

A successful infiltration requires time for planning; briefing; rehearsals; equipping; inspection; preliminary reconnaissance using maps, aerial photographs, and other sources; and any prepositioning that might be necessary. There are times when formal urban reconnaissance by divisional or Marine Expeditionary Force (MEF)–level assets is not feasible. LtGen Ron Christmas, USMC (Ret.), was a company commander during operations in Hue during the 1968 Tet offensive. He recalled that "when the battle began to retake the city, it was already a face-to-face fight. On 3 February, we fought our way into the [Military Assistance Command, Vietnam,] compound. When we took the front portion of what is now a five-star hotel, we were looking at the enemy right across the street" (Christmas, 2001). Tactical conditions and the time available denied then-Captain Christmas the benefits of more than the reconnaissance he could conduct by looking across a street from an upper story window.

A tension will inevitably arise between a maneuver commander's demands for rapid action and reconnaissance personnel's desire to fully prepare for and properly complete an assigned mission. When the decision is made to commit reconnaissance units, a thorough analysis of mission and environmental factors will be essential to determine how much time is available before other units initiate operations and which insertion method (or methods, because multiple insertions might be desirable) will best meet mission requirements. Despite the widespread acknowledgment of how difficult this task can be, a number of recommendations were put forth regarding how to increase the probability of success. Several have been used during operations. Unfortunately, from the perspective of this report's combat focus, most were during stability-related missions. They nonetheless offer prospective techniques that could have application during combat contingencies:

- "Maybe you can teach these guys how to be packaged . . . where the recon force gets delivered by someone else [and becomes a stay-behind force]. Maybe these guys get dropped off in a big box, like a Trojan Horse" (Paulovich, 2001f).
- Infiltrate during periods of reduced visibility if tactical conditions allow (though one Marine disagreed with this recommendation, stating that in some cases the hustle and bustle of day sometimes provides better cover) (Jirka, 2001).
- "The VC [Vietcong] infiltrated Saigon in small elements, including sapper units. They were well armed and equipped, possessing RPGs, AK-47s, mortars, and assorted new small arms. Each man . . . carried a basic load of ammunition as well as many grenades. Common techniques included stolen civilian, ARVN [Army of the Republic of Vietnam], and U.S. military vehicles to gain entrance into the city. Many of the VC operating independently throughout the city were dressed in police, ARVN Marine, or Ranger uniforms, as well as civilian clothing" ("Combat Operations After Action Report," 1968, p. 50).
- The "swarm technique": "Marines move mounted to a building they will occupy. The infantry rush out quickly into the building and clear it. After clearing, the troops return to the vehicles, except a few remain hidden in the building. Any OPFOR [opposing force] observing are unable to know exactly how many Marines entered the building and thus do not know that some

Marines remained behind. After the vehicles move away, the remaining Marines take up watch."[3] (It should be noted that at times the OPFOR would move into the building believing that the Marines would not return because the building had been cleared. The risk of this occurring must be assessed if the technique is being considered.)

- "Trying to blend in with locals was virtually impossible given the short hair, the way Americans carry themselves, and . . . other distinguishing characteristics. Their most successful infiltrations were done in conjunction with other forces in the area. An example comes from a USMC captain who worked with the British Army's Royal Dragoons. The Dragoons would go by a target building several times, halting there and elsewhere in the area. During one of the halts, likely at night, the [British] Marines would remain behind in the building and establish an OP. The problem was that civilians were constantly watching the NATO forces, children would play in the building, and avoiding compromise was therefore extremely difficult."[4]
- "The most feasible solution is to 'hide in broad daylight' (to be inserted via transportation means that makes the Marines look like other peacekeeping force troops) or to be dropped off for other functions (logistical or CMOC activities). Scouting teams can be 'packaged' for insertion by other groups (hidden in boxes, food convoys, etc.)" (Paulovich, 2001c, p. 3).
- "One potential innovation will be to overcome the difficulty in insert/extract by using an urban harbor site or patrol base to conduct mission planning and rest to enable multiple missions" (Paulovich, 2001c, p. 4).
- "We don't use partisan inserts enough [in which a native drives a team in or makes resupply runs]" (Morin, 2001; Jirka, 2001).[5]

---

[3]The source of this quotation, Col Fred Gellert (one of this report's authors), was a RAND Army Fellow at the time.

[4]Remarks by 2nd Reconnaissance Battalion representative at the March 2001 Urban Ground Reconnaissance Conference.

[5]"CMOC" is "Civil-Military Operations Center," a facility established to coordinate and monitor civilian and military operations in an area with the aim of mutually serving each organization's objectives.

- During Project Metropolis, one team stayed in a van virtually for an entire mission for several days and acted as an OP. A problem: the SINCGARS antenna had to be stuck out the window, increasing the chance that the team would be compromised (Jirka, 2001). (Another Marine suggested that the team could have disconnected the vehicle antenna from the vehicle and hooked it up for use with the SINCGARS system.) (Morin, 2001.)
- During Project Met[ropolis] a diversionary effort was conducted as follows: a helicopter was used to draw the enemy's attention while a recon team came in by foot or vehicle (Hardy, 2001).
- Colonel Tom Sward had a young officer who experimented with bicycles during the former's command of a Light Armored Vehicle (LAV) company. Sward noted that his Marines "easily strapped the bicycle to the side. . . . As you'd come up to a suspicious area you'd dismount. . . . We started using bicycles because it was much, much faster" (Sward, 2001).[6]

Two additional infiltration techniques are worthy of discussion—subterranean movement and the wearing of civilian attire. Though many requirements and not a few risks attend urban subterranean movement, the method has been used successfully in the past. The best currently available guidance regarding this approach comes from British doctrine. The level of detail and scope of tactical consideration provided in the relevant British Army Field Manual (AFM) merits extensive quotation here:

> The grenadier should carry the tools needed to open manhole covers. If the patrol is to move more than 200 to 300 meters [underground] or if the platoon commander directs, the grenadier should also carry a suitable telephone and line for communications. The point man should be equipped with night vision goggles, together with an infrared source, to maintain surveillance within the sewer. In addition, he requires a feeler for trip wires. All soldiers entering the sewer should carry a sketch of the sewer system to include magnetic north, azimuths, distances, and manholes. They should also carry protective masks, flashlights, gloves, and chalk for marking features along the route. The patrol should also

[6]Now-Maj Kelly Alexander was the platoon leader whose idea it was to use bicycles in this manner.

be equipped with a 120-foot safety rope to which each man is tied. To improve their footing in slippery sewers and storm drains, the members of the patrol could wrap chicken wire or screen wire around their boots. A constant concern to troops conducting a subterranean patrol is chemical defence. . . . Noxious gases from decomposing sewage, especially methane gas, can also pose a threat. These gases are not detected by the [nuclear, biological, chemical] detection system, nor are they completely filtered by the protective mask. Physical signs that indicate their presence in harmful quantities are nausea and dizziness. The patrol commander should be constantly alert to these signs and know the shortest route to the surface for fresh air.

. . . with the manhole cover removed. The patrol should wait 15 minutes before entry to allow any gases to dissipate. The point man should remain in the tunnel for 10 minutes before the rest of the patrol follows. If he becomes ill or is exposed to danger, he can be pulled out by the safety rope. When the patrol is moving through the tunnel, the point man moves about 10 meters in front of the patrol commander. Other patrol members maintain five-meter intervals. If the water in the tunnel is flowing fast or if the sewer contains slippery obstacles, those intervals should be increased to prevent all patrol members from falling if one man slips.

The confined space of tunnels and sewers amplifies the sounds of weapons firing to a dangerous level. The overpressure from grenades and mines exploding in a sewer or tunnel can have adverse effects on friendly troops such as ruptured eardrums and wounds from flying debris. Also gases found in sewers can be ignited by the blast effect of these munitions. For these reasons, small arms weapons should be the principal weapons systems employed in tunnels and sewers. . . . Physical and mental fitness can be maintained by periodically rotating soldiers away from tunnels so they can stand and walk in fresh air and sunlight. Stress management is also a factor of operations in tunnels. Historically, combat in built-up areas has been one of the most stressful forms of combat. Continuous darkness and restricted maneuver space cause more stress to soldiers than street fighting." (Army Field Manual, Volume IV, 1998, pp. 4-2–4-4.)[7]

---

[7]For more on subterranean operations, refer to U.S. Army, 2001, pp. 9-32–9-34.

The 1999 version of the same manual further notes that "there will be a requirement for escape routes in case of entrapment by the enemy or collapse of the subterranean system. . . . Reconnaissance of the subterranean passages of a platoon or company area should normally be the responsibility of a patrol of section [squad] size. Only in extremely large subterranean features should the size of a patrol be increased" (Army Field Manual, Volume 2, 1999, p. 42).

The use of subterranean passages (including subway tunnels, utility pipes, and others in addition to the sewer systems addressed in the AFM) has been considered by USMC units and is evident in the following comments by two members of the 1st Force Reconnaissance Company. Equally evident is the requirement to develop more extensive doctrine and related training to guide and practice relevant TTP:

> They say go through sewers, but they don't tell you about poisonous gasses, what immunizations are needed, how to navigate, what obstacles you'll come across. GPS doesn't work in sewers. The unit I know of that was confronted with such a situation asked for guidance and [they] were told just to move out and execute getting from point A to B. . . . We need to know more about any subways in the city. What would we in general find there? How are electric different from diesel? How could a unit use exhaust vents or other features of the system to its advantage?[8]

The comment regarding GPS is notable. Underground navigation is notoriously difficult. Use of a compass and pace count may be of value in some instances when a map of the system is at hand, but the presence of electromagnetic sources in or near many subterranean systems suggests that a team must also be prepared to employ alternative methods. Development of effective inertial navigation devices that allow the user to program in a start point and determine his location based on subsequent direction and distance traveled may not be long in coming. Until then, units must rely on field-expedient methods and any available system maps (e.g., subway maps or city maps showing subway station locations). It may be possible to augment maps and sketches with overhead photography that shows

---

[8]1st Force Reconnaissance Company interviews.

surface signs of subterranean passageways (such as manhole covers) or special photography that reflects electromagnetic or other signatures characteristic of specific types of underground passageways.

Finally, discussion of infiltration techniques would be incomplete were it not to include consideration of reconnaissance Marines wearing civilian attire to improve their chances of remaining undetected. If Marines are given the option of employing such a technique, the nature of the urban area and the conditions extant at the time of infiltration will be elemental in their influence. U.S. special operations forces' use of native garb in urban and rural Afghan environments proved itself viable. The extent that such garb covers both the body and face significantly abetted this success. Wearing local civilian garb in environments in which less is covered may prove to be of little advantage, especially if the military personnel need to carry weapons and large items of equipment when either or both is not the norm. In metropolitan urban areas, it may be feasible for Marines to wear concealing clothing different from what is worn by most indigenous personnel but similar to that used by minority segments of the population or visitors. Concealment should not be limited to clothing. Wagons, bicycles, racks for carrying large loads on the back, and other local items may both aid in concealing the character of the user and in covering materials that must accompany a reconnaissance team. (The use of civilian clothing and interpretation of Marines' status as combatants should they be captured have implications. The character of the enemy and the manner in which the civilian clothes are worn [e.g., are they mixed with legitimate uniform parts?] may influence a commander's decision whether or not to permit this procedure. The latter may have an effect on the status of those wearing the civilian attire. Situation-specific questions should be directed to military legal experts so that the potential consequences of such decisions are known.)

Urban Reconnaissance Exfiltration, Escape, or Evasion

> It's easy when you're hunkered down in a spider hole and they're only coming from one direction. Now they're coming from all directions.
>
> Col Michael J. Paulovich, USMC

Many of the same considerations that apply to infiltration efforts pertain in equal measure to exfiltration during patrolling, from a hide, or from other reconnaissance positions. Related tactics, techniques, and procedures will therefore apply in similar measure. The situation is somewhat different when a unit is detected, whether during a failed infiltration, compromised exfiltration, or discovery of a hide position. In such cases, the appropriate reaction can range from quick thinking and a hasty verbal response to escape and evasion under fire. British reconnaissance personnel demonstrated a consciousness of local concerns and not a little savvy when operating in Bosnia. If compromised, they would say they were trying to catch car thieves. The locals were supportive because vehicles were regularly being stolen. The military personnel were actually looking for Serb leaders and monitoring Serb actions (Kerkering, 2001). Unfortunately, such adroit responses will have little value in most combat situations.

The aforementioned discussion of short-range engagements and limited sight distance makes it evident that a compromised unit may well find itself with very limited routes of egress when confronted by an enemy within 100 meters of its position. Appropriate responses could include using large amounts of suppressive direct fire to intimidate the compromising force, hunkering down and calling for a rapid reaction force with a short response time, or initiating a fire support plan with the capability to strike targets in very close proximity to the endangered team. If breaking contact is called for, the withdrawal may involve a continuous movement back to an area controlled by friendly forces, closing on a vehicle or air extraction point, or the use of intermediate locations. The last may be rally points at which several members of the team rejoin, or it may be one or more safe houses that provide a place to conceal themselves, rest before continuing a long movement, or render aid to wounded or injured.[9] Several Marines interviewed expressed the desire to receive instruction on hot-wiring vehicles for use during urban movements, especially emergency exfiltrations.

The closer proximity of friendly forces to enemy and noncombatant personnel will fundamentally influence the timing and tactics

---

[9]The concept of using safe houses comes from Root (2000).

involved in given missions. Open terrain that allows long-range engagement by Marine snipers, indirect weapons, or other means generally provides flexibility sufficient to render withdrawal not especially problematic after the contact. It may still be feasible for the reconnaissance or surveillance asset to remain in place after the event. The same will seldom be true in a town or city, even in cases in which Marines employ sound-suppressed weapons. The proximity of enemy and friendly personnel that is the norm means that direct fire engagement locations are likely to be rapidly compromised or that the number of possible hides makes subsequent compromise very likely. Dusk or evening engagements may be desirable to facilitate escape after engagement (Ziegler, 2001, p. 1).

Escape and evasion plans and those for coordinating fire support should be coordinated and made as simple as possible for both reconnaissance personnel and those supporting them. Marines with experience in the field decried the frequent changes of fire support plans that occurred during operations. They also noted that the several organizations that might assist during an escape and evasion procedure too often failed to coordinate their plans with each other. This requires deployed reconnaissance units to prepare for separate procedures for assistance rendered by wing-level organizations, a battalion landing team, or some other unit. Ideally the extraction procedures and plans would be identical. At a minimum, they ought to be coordinated.

One remaining point is worthy of note. Intelligence services have been more willing to provide some products in a timely fashion to tactical units in recent years. Less than five years ago, many such units found it difficult to obtain overhead photography images because of agency concerns regarding compromise of the source. As a result, lives were put at risk as patrols attempted to negotiate urban terrain with small-scale maps or tourist graphics. Cooperation in this area has subsequently much improved. There unfortunately remain instances in which some intelligence organization personnel continue to withhold information vital to operational success and personal survivability in instances where greater wisdom in judgment and expedience in release could mean the difference between life and death.

Urban Maneuver and Target Engagement

> You can't stay on the street. If you stay on the street you're dead. You have to go through buildings and walls.
>
> LtGen Ernest C. Cheatham, USMC (Ret.)

> Normally we fired on guesses. . . . We had very few observed artillery fire missions.
>
> BGen Michael Downs, USMC (Ret.)

Reconnaissance and surveillance personnel should be trained in all aspects of urban operations maneuver and related tactics, techniques, and procedures. Though their mission will often call for avoiding detection by the enemy, the density of adversary personnel in built-up areas and the resultant increased proximity of reconnaissance to adversary forces mean that inadvertent contact will be more common than on other terrain. Regardless of whether their purpose is offensive or defensive, team members must be capable of safely entering and clearing rooms and buildings, moving between and within structures, calling for fires, establishing communications, evacuating wounded, and performing the many other tasks that require special training or adaptation when executed in an urban environment. As one Marine observed, "a team may enter a building to set up an OP, FFP [Final Firing Position], etc., and still have to clear it just as an infantry squad has to. The team cannot afford to not clear the rooms and have enemy personnel within the building" (Ziegler, 2001).

Similarly, reconnaissance personnel will require knowledge of how to select and prepare equipment appropriate for their missions. Carbines (M4s rather than M16s) will facilitate urban movement. Soft-soled boots and movement techniques designed to minimize noise or contact with hard surfaces will reduce chances of compromise during displacements. Zeroing individual weapons for engagements of 50 meters rather than in excess of 100 meters will be appropriate for most if not all members of a team (Kerkering, 2001). Smoke reacts differently in streets than in the often swirling air of intersections. Tamping explosive charges so that they are not heard beyond a range of tens of meters may be sufficient for rural targets. Tens if not hundreds of people may feel or hear the same detonation as it shakes the structure in which they reside or as its report reverberates off building walls. Team members will ideally have a nonlethal means of silencing a noncombatant who accidentally stumbles on an in-progress mission. If they lack such means, they must have a drill for dealing with these situations. They also must practice rapid reaction drills uncommonly called for elsewhere—e.g., quickly or preemptively silencing a barking dog.

Medical Evacuation and In-Place Medical Treatment of Wounded or Injured[10]

> I suppose one of the things that we learned very quickly in Hue City was that if a man was wounded, you didn't pull back and call for a corpsman because then you'd have two guys down. . . . We found out that in the city you have to carry on and leave they guy, that the guy had a better chance of surviving if we just kept going and let the guys behind us pick him up. It's a difficult thing to do.
>
> LtGen Ernest C. Cheatham, USMC (Ret.)

The treatment of wounds and injuries incurred during urban operations and the evacuation of patients can be far more difficult than is elsewhere the case. The task can be especially demanding if the number of personnel available to evacuate the downed warrior and provide covering fire comes from a four- or six-man reconnaissance team. Military personnel confronted with such situations in the past have found it impossible to both treat the wounded and maintain the level of fire requisite for unit survival. Speed of emergency treatment or Marine self-aid will both tend to mitigate resulting problems. The U.S. Army medical community continues in its efforts to test and obtain approval for the acquisition of hemostatic dressings, as well as a one-handed tourniquet. Both Marine and Army units have expressed interest in QuikClot, a rapid clotting agent that has been found effective in treating severely bleeding wounds or injuries (MCWL, 2002). These and other advances will help to address some

---

[10]The RAND Urban Operations Team is also working on an urban logistics study for the U.S. Army Combined Arms Support Command (CASCOM) at the time of this writing. Analysis will include extensive consideration of urban medical operations. The final report will be available in early 2003.

of these challenges by the end of calendar year 2002.[11] Increasing the medical expertise of reconnaissance personnel may also be beneficial, including instruction on such tasks as stitching a wound. Gashes and cuts occurred with greater frequency than normal during 1st Reconnaissance Battalion participation in Operation Metropolis (Hinton, 2001).

Urban Aviation and Air Defense–Related Tasks

> The idea of a helicopter-borne urban reconnaissance just scares me to death. . . . Unless you can find a helicopter that is silent, it's dead.
>
> LtGen Ernest C. Cheatham, USMC (Ret.)

A comprehensive study of urbanization's effects on rotary-wing and fixed-wing operations is as yet unavailable. The urban ground reconnaissance focus of this effort precludes too great a discussion in that regard, but several factors are noteworthy, given the importance of this support to Marines in less restricted terrain. Aviation can be especially vulnerable over built-up areas, as was demonstrated in Mogadishu and Grozny. Pending adaptive tactics and technologies, this vulnerability is very likely to influence commanders' willingness to commit helicopters to many of the tasks they routinely undertake on other ground. Vulnerability is only part of the issue. Engagement ranges of less than 1,000 meters render some types of "smart" munitions useless because the distance required for the weapons to arm or acquire guidance mechanisms (i.e., lasers) is too great. Helicopters employed in such scenarios will therefore likely need a mix of smart and "dumb" munitions. The options available may not be optimum. Cobra 2.75-inch rockets can be too greatly affected by rotor wash, urban winds, or drafts created by fires to risk use in proximity to friendly forces, noncombatants, or other proscribed targets.[12]

---

[11]The U.S. Army is pursuing projects to field one-handed tourniquets and hemostatic bandages during 2002. The latter cost $300 each and have yet to be approved by the Food and Drug Administration, but the service had established a goal of fielding at least 1,000 of the bandages for operational use by September 2002. The goal was to also field 1,000 one-handed tourniquets (approximately the size of the current first-aid pouch issued to each Marine and costing $10 to $15 each) by the end of the year (Modrow, 2002).

[12]For a description of Cobra munitions concerns and effects during 10th Mountain Division (U.S. Army) operations in Mogadishu, see Casper, 2001, pp. 69–70.

Marines tend to use their helicopters in a close air support (CAS) role much more frequently than do their Army counterparts (and their AH-1s might therefore be allocated to support reconnaissance E&E or other missions). Therefore, it would be advisable to give consideration to increasing the quantity of 20-mm cannon ammunition and ensuring that all aircraft have the equipment necessary for limited-visibility operations.

A symbiotic relationship may exist between ground urban reconnaissance elements and aviation assets. Although findings are preliminary, some participants in the Victorville experiment observed that ground forces were more effective at detecting and eliminating enemy air defense threats in a built-up area than were those in the air. A drawback was that any ground element executing such a task would be compromised. Because reconnaissance assets would likely be dedicated to supporting air insertions, extractions, or other aviation actions, there is a call for developing effective TTP for detection, monitoring, and engagement of the foe's air defense capabilities and ground reconnaissance roles in support of those undertakings. It is similarly apparent that reconnaissance personnel should be trained in selecting landing zones (LZs), supply drop zones (DZs), target approaches, attack positions, target designation points, and access and egress routes for aircraft in support of urban missions. Such training should include implications of limited visibility aviation operations, including those at night when lights, fires, smoke, and other elements can influence aircraft safety and pilot vision.

RECONNAISSANCE AS A SYSTEM AND SUBSYSTEM

Introduction to the Second Theme

Second Theme. Tactical ground reconnaissance is a system of systems within a system.

> Destroying, controlling, or protecting vital parts of the infrastructure can isolate the threat from potential sources of support. Because these systems are inextricably linked, destroying or disrupting any portion of the urban infrastructure can [also] have cascading effects (either intentional or unintentional) on the other elements of the infrastructure.
>
> FM 3-20.96, *RSTA Squadron* (Second Coordinating Draft)

> The study of the objective must provide a complete picture of the enemy's firing points and fire system. It must produce complete familiarity with the roads of approach and enable the commander to estimate the best time for the storm. Reconnaissance also must give information on the nature of the defenses, the thickness of walls and obstacles, the position of entrances, concealed loopholes and [ways] of communication, on the directional sectors covered by enemy fire, on the obstacles in front of strong points, and on the position of the firing points of neighboring strong points which can keep the approaches under flanking fire. If necessary, the information supplied by reconnaissance must be checked up by battle patrol.
>
> V. I. Chuykov, "Tactics of Street Fighting: Military Lessons of Stalingrad"

Marine ground reconnaissance is always a system of systems that is itself a vital component of a larger system. Reconnaissance provides intelligence of value in planning subsequent operations, and it continues to provide input during the execution and extension of those undertakings. The urban environment does not change this fundamental truth, but it does impose special requirements both of scope and type of work. The closeness of the environment and related increased density of mission-relevant participants means that Marines may be collecting for (and from) PVOs; NGOs; coalition members; and indigenous police, military, or civil authorities in addition to more traditional customers. They may also be doing so in far greater volume than is the case elsewhere. The problem is exacerbated by the quantity of information that an urban area surrenders even without the additional potential users. As alluded to earlier in this report, intelligence units and staff sections from the sharp end through MEU and even Marine Expeditionary Brigade (MEB) headquarters levels may lack the capacity to synthesize the input received barring modification of normal precedence and reporting procedures. This is especially true if commanders compensate for increased urban densities and restricted LOS by putting more reconnaissance collectors in the field.

Marine Urban Reconnaissance as a Subsystem

> Doctrine states that reconnaissance is required to reveal 70–80 percent of targets. The Chechens, who were deployed in small, mobile

> groupings, made the job of locating them very difficult. That said, the Russian advance and initial assault illustrated a clear conflict in the doctrine between the need for speed and the requirements for effective reconnaissance. . . . The relentless requirements of Russian doctrine for speed were unobtainable and clashed with a need for effective reconnaissance. This resulted in a botched assault, which was exploited by the Chechens.
>
> British Army Field Manual, 5-1

> We can look at a 1:50,000 map [on other terrain] and know what the terrain looks like. . . . Guys got lost in Mogadishu, but if you put those guys in downtown Hawthorne, California, they'd get lost there too.
>
> Col Michael J. Paulovich, USMC

Urban reconnaissance is foremost a component in the larger system that is (for the purposes of this study) Marine combat forces and the coalitions of which they are a part. Ultimately the responsibilities inherent in this status should dictate the allocation of duties between echelons within the Marine reconnaissance community just as is the case in any environment. Because distances are compressed in and around urban areas, those reconnaissance assets that traditionally are distant from main force units may be much closer. The short distances characteristic of urban warfare could mean that the same point serves as both NAI and TAI. This further taxes the reconnaissance element observing that location. It now has the additional responsibilities of determining likely or confirmed enemy air defense capabilities in the vicinity, potential ground force ambush positions, and feasible flight paths for indirect fire munitions to avoid inadvertently striking tall buildings between the target and guns.

This last point highlights the need for reconnaissance teams to have an even more comprehensive knowledge of friendly force equipment than is otherwise the case. Vehicle weight, width, and height are essential knowledge for the reconnaissance Marine determining routes for friendly units. Urban operations demand other considerations rarely a concern elsewhere. These include the following:

- Elevation, depression, and traverse limits on guns to determine whether a column will be vulnerable while on a given route.

- Condition of structures along the route. Vibration from heavy vehicles can cause the collapse of poorly constructed or battle-damaged buildings, presenting a risk to friendly force and non-combatant personnel.
- Weapon systems–unique concerns. In addition to problems with round trajectory, artillery and mortar units must be positioned to avoid tightly enclosed areas, structures that may collapse beneath them after repeated firings, or areas for which resupply is problematic.
- Range, arming, and impact characteristics of various munitions. Positions for vehicles providing direct fire support should capitalize on maximum effective weapons ranges when possible (Army Field Manual, Volume 2, 1999, p. 2-8). The previously mentioned concern about the distance necessary for helicopter-launched missiles to arm similarly affects weapons even at the lowest unit levels (LAW, AT-4, and M203 grenades). A reconnaissance team should be able to determine and report the apparent nature of building materials in structures that might have to be engaged so attacking units can appropriately mix weapons systems and ammunition. Training rounds [e.g., training practice tracer (TPT) for the Bradley 25-mm gun] have in some cases been found more effective for penetrating walls than high explosive (too much lateral effect) or sabot (tend to punch straight through some materials creating nothing more than a small hole).

It has already been mentioned that the urban environment's greater density of activities, forces, HUMINT sources, and tactically significant terrain features combine with limited LOS to increase the volume of information reported and can promote a need for deployment of more reconnaissance assets at any given time. The potential for the resulting deluge of incoming intelligence to overwhelm tactical-level headquarters' handling capabilities has also been discussed. Potential solutions at the headquarters include increasing the manning strength of intelligence staff sections, enhancing and enforcing procedures for assigning precedence to incoming ***information, and more detailed Commander's Critical***

Information Requirements (CCIR).[13] The challenges can also be addressed from the reporting node perspective. Teams can be more discriminate in what they choose to report. (The implications for ensuring that reconnaissance personnel are kept well informed with respect to changes in mission and commander's intent are obvious.) Assignment of an intelligence synthesizer to each team or group of teams would insert an additional node between the reporting units and receiving headquarters. This individual would accompany reconnaissance units during missions. He would be capable of filtering incoming information in light of his personal grasp of the situation and including only what is necessary when forwarding his reports.

A form of this was employed during the Project Metropolis experiment at Victorville, though it was line infantry squads rather than reconnaissance units that employed the asset. An experienced noncommissioned officer (NCO), generally a former squad leader, was the "information manager" for a squad. All incoming reports from squad members went through him, and he, using his personal situational awareness and savvy born of experience, forwarded only what he thought of value to higher headquarters. The process was thought to be effective, but the obvious limitation is that too few Marines have such experience and those who are with units are likely needed for other responsibilities. Demanding more analysis from reporting teams is also an option. Rather than submitting an observation regarding the siting of "three T72 tanks and one BMP," reconnaissance personnel could be asked to further provide an initial prognostication as to the character, size, and intentions of the enemy force involved. This could have notable pertinence during urban contingencies when the adversary might use several parallel streets, multiple floors, or otherwise disperse his units so that a reconnaissance element would see but a fraction of the whole.

A well-trained reconnaissance unit might be able to perform considerable analysis based on the types of systems seen (e.g., antitank weapons could signal that the personnel or vehicles spotted were on the flanks of a larger moving body), others heard, and further clues as

[13]Brian Ziegler (2001, p. 4) notes that "an R&S team has to be able to maintain an information log and decipher what needs to be reported now and what can wait," sending only what is necessary and doing so with an appropriate message priority.

to how the fragments of information fit into the larger situation. The result would be much-condensed reports that reflect intelligence of greater immediate value than raw reporting and therefore are available for more rapid dissemination to other units. This timeliness of reporting is significant in an environment where distances between reconnaissance units and maneuver elements are limited. Reconnaissance team personnel are often on average more experienced than many other types of units, and that lends further credence to a conclusion that such first-line analysis would generally be of high quality.[14] The shortfalls in this approach include reliance on experience and good judgment: the demands on the training system, senior unit leadership, and personnel system (especially were the Marine Corps to undergo a rapid, Vietnam-type expansion) are considerable.

Perhaps the way of most readily addressing the greater demands for urban reconnaissance and surveillance is better integration of all potential information collectors. Reconnaissance units are but a part of a larger collection system that includes infantry patrols, civil affairs units, fixed-wing and helicopter personnel, and support Marines in the AO, to name just a few. These assets are sometimes capitalized upon, but they are rarely used to full capacity. Colonel Michael J. Paulovich, while commander of the 1st Reconnaissance Battalion, sent Marines out to share a cup of coffee with comrades in other units, asking what truck drivers, engineers, and others observed while performing their duties. They collected intelligence through a shadow system to complement what the battalion's "formal" reconnaissance effort was finding. Ambassador Robert W. Farrand used a similar procedure to increase his situational awareness while serving as Deputy High Commissioner and Ambassador in Brcko, Bosnia-Herzegovina. He would frequent coffee shops and strike up conversations with local nationals. Those responsible for the collection and synthesis of intelligence should work with subordinate, peer, and senior headquarters as well as other organizations to ensure that all relevant military personnel are educated with regard to intelligence requirements, encouraged to report information of value, and regularly debriefed after missions.

---

[14]The authors thank Col Pat Garrett for his observation regarding the experience level of reconnaissance NCOs (Garrett, 2001).

A commander must recognize that virtually any of his assets, as well as those of other service, joint, coalition, PVO, NGO, and indigenous source organizations or individuals, can be sources of intelligence. He should properly train, equip, and debrief sentries, road guards, and other security personnel. John Allison, who was responsible for force protection in 1992–1993 Mogadishu, recalled that sentries "aren't recognized as [sources of information]. We don't resource those guys with jack. We need to give them better night vision goggles, radios. If we made them feel more a part of the mission, we'd get a lot more out of them. . . . Beirut was an intelligence and reconnaissance failure. . . . If we'd trained guys down below to look for what they should have been looking for, it might never have happened" (Allison, 2001).

The urban environment makes it especially difficult to determine the most effective locations for OPs and the best routes for reconnaissance patrols. Overflights are themselves valuable for collection, but leaders should not overlook the planning value of such assets. While in Mogadishu, John Allison would use both rotary-wing and fixed-wing assets to "look down from above, because you can see things from above that you can't see from the ground. . . . We used the AC-130 in 1995 for intel and presence. Because it had all the night vision, we could see everything that was going on" (Allison, 2001).

Tactical-level Marine reconnaissance assets will make unusual demands on mapping and overhead photography capabilities when the mission takes them into urban environments. The map and imagery scales needed will be much larger than is the standard. The Israelis, like the British in Northern Ireland and U.S. Army in Berlin, found that even 1:25,000 maps are of little value. Brigadier General Gadi Shamni (IDF), Head of Israeli Defense Force Infantry and Paratrooper Doctrine, noted that his army routinely uses 1:2,500 maps in urban areas. "On these maps you have areas that will be 1:500, and these maps will have items that allow you to communicate, and we keep updating these maps" (Shamni, 2002). The use of such maps comes from lessons learned at great cost. One of the reasons the Israeli Army had such problems in 1973 Suez City was that the battalions and brigade attacking into the built-up area had only 1:100,000 maps as they moved directly from their operations in the desert. Fortunately, the territorial command headquarters had extremely detailed aerial photographs. Personnel on the command staff bypassed the division and brigade echelons to work with the S3 of the second infantry battalion in planning the exfiltration from the city. The brigade commander recalled that the 1:100,000 scale maps "were useless for urban navigation" (Karen, 2002).[15]

British doctrine makes the further observation that "more information can be gained by using aerial and satellite photographs and maps together rather than using either one alone" (Army Field Manual, Volume IV, 1998. p. 3-30). While progress has been made in providing tactical units with previously close-hold overhead imagery, the synchronization, synthesis, and distribution of imagery still greatly lags behind tactical needs. The recognition is to some extent there; the provision of UAV images directly to pilots when the unmanned systems detect air defense systems has begun. However, intelligence command and control systems that can rapidly respond to requests for images in the field, select and task the appropriate system, and return a usable product in a timely manner are too rare. Immediate and comprehensive action with regard to relevant doctrine, systems acquisition, training, and organizational procedures is long overdue in the U.S. armed services. The problem is often one of upper echelon dissemination policy and an inability to timely get information to the user. Those suffering the consequences of inaction or malfeasance are the men and women at the cutting edge.

Marine Urban Reconnaissance as a System

Marine Corps tactical ground reconnaissance has several subsystems in addition to its being a system itself. Those of particular interest to this investigation are STA teams, division reconnaissance battalions, and the force reconnaissance company. Surveillance and target acquisition teams are infantry battalion assets. Depending on the situation, their responsibilities might include those of sniper; reconnaissance; fixed-point or convoy security; forward observer for ground, air, or sea fires; or very specific missions, such as eliminating enemy urban air defense capabilities. Of the three types of reconnaissance elements, STA teams are those likely to be closest to regular maneuver unit positions. Next in the topographic tier are division

---

[15]General Karen was brigade commander of 500 Armor Brigade, 162 Armor Division, during the battle of Suez City, October 1973.

reconnaissance battalion units, with force reconnaissance teams traditionally being the most distant. The specific distances will differ, depending on terrain, mission, and other factors in addition to supporting fires. Urban terrain and related missions will likely put these and other types of units closer to each other than is the norm elsewhere.

Distance is important, but reporting responsibilities will have even greater impact on the timeliness of intelligence dissemination. STA teams obviously provide a MEU commander with an asset that reports directly to his own headquarters. Their information should therefore suffer little delay in getting to Marines with the companies, platoons, and squads that are farthest forward. Divisional reconnaissance intelligence, having to go through the filter of at least one additional headquarters, is somewhat delayed, barring the establishment of a special reporting/monitoring relationship between the MEU commander and higher-echelon leaders or staff.

The passage of intelligence from force reconnaissance assets to forward-most units can take considerable time. MEU-based Marines are only half joking when they complain that they never see the benefits of force reconnaissance work but suffer when the teams are compromised and the MEU has to assist in their recovery. The dangers of delay are potentially greater in built-up areas because each headquarters will have a greater volume of information to synthesize and distribute. The closer proximity of these various assets and the potentially more immediate consequences of not getting vital information to the appropriate unit in a timely fashion dictate that modification of current reporting procedures be considered. Division and force reconnaissance assets could include specified intelligence users in given reports or the provision that those users be allowed to (and be equipped to) monitor relevant reports. The model for determining which users ought to monitor what higher-echelon reconnaissance asset reports could emulate that for NAI and TAI: Links between certain NAIs (which could be topographic or activity-based in character) and specified units would determine what reports go to which users. The means of providing this intelligence could be frequency monitoring, e-mail, instant messaging, or any other mechanism provided that all parties possess the requisite equipment, manning, and training to establish such a system.

Several currently controversial topics have relevance in determining the appropriate roles for USMC reconnaissance assets during urban contingencies. First, there is a long-standing debate about whether STA team members should be primarily "shooters" (snipers) or surveillance assets. As a reconnaissance battalion commander, Colonel Paulovich found that "the sniper culture works heavily on students to teach them that they are stalkers and hunters. I sent as many as I could to Scout Sniper School because the crossover skills of stalking, observation, [and establishing] hides were great. I had to deprogram them upon return to get out of the shooting role" (Paulovich, 2001d). He went on to further explain why the surveillance duties were important: "I'll echo what the battalion commanders I talked to on this [said]. They don't want to be blind, especially [during] urban [operations]. Our Somalia experience . . . was that most of our sniping engagements were at closer range and most Marines who were designated marksmen were doing quite well. Snipers are great to have, but good intel [is even more important]" (Paulovich, 2001d).

The U.S. Army's LTC Dave Maxwell, commanding the Special Forces battalion conducting operations in the Philippines during 2002, agreed that both the killer and reconnaissance tasks are vital, as do those designing that service's Interim Brigade Combat Team (IBCT) at Fort Lewis, Washington. Sergeant First Class Tommy Hooten, the NCO in charge of the U.S. Army Sniper School at Fort Benning, Ga., concurs: "Marksmanship is a big part of it, but it's also about reconnaissance. . . . In some cases, that sniper is the only eyes and ears that the commander has on the ground." Hooten further believes that the psychological impact of having snipers at checkpoints as a show of force and visible on rooftops to intimidate potential troublemakers during peacekeeping missions is important (Cox, 2001, p. 19).[16] The Israelis disagreed, seeing the primary sniper mission as one of shooter. General Shamni summarized the perspective, "We see reconnaissance as someone who moves. A sniper is stationary. He is not a maneuverable element. We see reconnaissance as a maneu-

---

[16]The article also notes that the U.S. Army's "Interim Brigade Combat Teams . . . are slated to include two sniper teams for each battalion. Each company would also have one sniper team" (Cox, 2001, p. 18). The snipers are also shown on IBCT force structure charts received by author Russell W. Glenn during interviews conducted at Fort Lewis, Wash.

verable element, someone who can give you the tactical meaning of the situation. . . . The only one that can give you this information is reconnaissance. . . . Reconnaissance is not the sniper's mission" (Shamni, 2002).

The trade-offs at the tactical level were very effectively summarized by a captain participating in the MCWL's Project Metropolis. This officer found that his greatest problem regarding the employment of STA teams was "differentiating between being a shooter and a collector. Snipers wanted to shoot all targets that they observed, despite the fact that this could have compromised their presence and possibly led to their being denied the opportunity to continue operations as a collection asset. During some operations, the teams were there to gain and maintain harassing fire with enemy forces. Other times they were tasked to collect information and were under restrictive fire controls" (Mangan, 2001).

The answer seems straightforward; the implications are not. Ideally, those assigned to STA teams should be equally adept at surveillance, sniper tasks, calling for fire, and others as necessary in a given theater for specific operations. The primary pragmatic considerations are serious ones: Multitasking demands more training time and therefore training dollars and some would offer that STA team personnel cannot be expected to execute many tasks as well as a single one. The costs are real, but an urban environment's voracious demand for reconnaissance assets goes a long way toward justifying the additional time and money. Those costs are likely not as great as might be originally expected. As noted by Colonel Paulovich, many shooter skills will serve well during surveillance, forward observer, or other potential STA team missions. The same overlap acts to reduce dangers that team personnel will be unable to master all necessary skills. There is little reason a STA team cannot perform as a reconnaissance asset as effectively as it can as a direct action resource. In truth, training for multiple tasks may well provide the teams with complementary skills that make them better at all tasks.

A second roles and missions debate concerns Marine Corps force reconnaissance companies. During deployments, a platoon from this organization normally accompanies a Marine Air-Ground Task Force (MAGTF). The preliminary training that this asset receives focuses almost entirely on direct action missions. The unit will fire thousands of rounds during its six months of SOTG training while spending little time on reconnaissance skills. As previously noted, the emphasis is such that many personnel in the reconnaissance community decry the atrophy of these skills suffered during the time spent in the SOTG environment. Similar to the debate regarding STA teams, it is legitimate to ask whether this seemingly lopsided emphasis is appropriate for personnel likely to see commitment to urban contingencies.

The concern encompasses more than loss of skills. When deployed, the force reconnaissance platoon may be the only direct action, in-theater capability allocated to a MAGTF commander, JTF commander, or combatant commander. It is often harbored as a de facto reserve force, kept aboard ship rather than committed to reconnaissance tasks. When a direct action mission does arise, it generally goes to special operations forces specifically trained for such an undertaking (e.g., Delta Force, SEAL teams, or Ranger elements) and brought in when needed. The opportunity cost of lacking the highly trained Marines in the field as reconnaissance assets will be felt all the more in an urban area where such resources are very likely already at a premium. In addition, stripping this layer from the USMC reconnaissance system means that a commander may have to forgo deep ground reconnaissance altogether. "Deep" in a city may be but several blocks instead of 50 kilometers or 50 miles. The impact of this loss may be felt more severely in the urban environment, where even operational-level reserve forces might be but a few kilometers and less than an hour distant.

The problem is a more complicated one in the force reconnaissance case. The extensive time necessary to obtain proficiency in the many facets of various direct action missions likely precludes gaining and maintaining acceptable proficiency in both reconnaissance and the entire spectrum of direct action skills. The question is one of "Which should take precedence?" Some argue that the situation cannot be altered, that the Marine Corps has essentially promised combatant commanders a direct action capability, and that these leaders will be loath to surrender them. One Marine put it succinctly: "You can't [eliminate the direct action role]; it's a political thing. . . . It even goes outside of the Marine Corps. We've sold certain things to the [combatant commander]. . . . We need to deemphasize the shooter mission, but I don't think we can get rid of it altogether." Others point out that the force reconnaissance community has its share of leaders that still believe their units will never be committed to urban contingencies. A 1999 review of MEU(SOC) missions resulted in no change to the force reconnaissance direct action orientation, lending credence to the belief that it may have achieved the status of a sacred cow.

Considering the question on its operational merits alone, it is first interesting to note that the primacy of the direct action mission is relatively new. Force reconnaissance units were viewed as the USMC's premier deep reconnaissance assets during the Vietnam War. They were neither equipped nor trained for direct action missions. The causes behind the subsequent reorientation are unimportant here. What is significant is whether this concentration of highly skilled and experienced Marines is being appropriately employed. The argument regarding the shortage of reconnaissance assets during urban operations obviously supports reduction or elimination of the direct action role.

While there may be instances in which a commander needs to conduct such missions with only the capabilities on hand, such cases would be rare. Major urban undertakings, even during operations involving extended combat, would merit commitment of special operations forces better equipped and trained for these tasks. Even were Marines the only asset in theater, the rapid deployment capability of such special operators means that other resources could quickly be at hand (and, in fact, they would likely have access to better preparatory training and rehearsal facilities before mission execution than would in-theater assets). A reorientation of force reconnaissance missions therefore seems feasible without undue risk to operational readiness.[17] A second alternative is to modify the direct action requirements to make proficiency in both that arena and reconnaissance feasible. An acceptable balance between the two skill areas might be attainable by limiting direct action training (and related equipping and other factors) to only those tasks deemed most

[17]It is worthy of note that special operations forces brought into a theater could remain under the control of Special Operations Command rather than the combatant commander or JTF commander with responsibility in the theater. This does have potential operational effectiveness implications (as was demonstrated in the negative sense on October 3–4, 1993, in Mogadishu, Somalia) as well as political implications.

likely to be needed or others that would abet a smooth handover to special operations assets.

Interservice exchange of intelligence continues to be a problem despite initiatives and dictates to redress long-standing issues. Yet, there is room for optimism. The recent information-sharing agreement between the Marine Corps and Special Operations Command "to better coordinate global operations of the two military components in the future" holds potential for improved intelligence feeds and information exchanges that could benefit both parties (Crawley, 2002, pp. 18–19).

LEADERSHIP, TRAINING, STRUCTURE, AND MANAGEMENT IN LIGHT OF URBAN RECONNAISSANCE CHALLENGES

Introduction to the Third Theme

Third Theme. Urban operations impose extraordinary leadership, training, task organization, and personnel management demands.

> One of your biggest problems in urban warfare was control. As a battalion commander my control was simple. The company commander, his control was more difficult than mine. But the guy that really had the problem was the squad leader, and there we're talking about the reconnaissance team.
>
> LtGen Ernest C. Cheatham, USMC (Ret.)

Leadership

> We were beat. Your mind, as a leader, was intensely engaged. You never rest.
>
> BGen Michael P. Downs, USMC (Ret.)

A need for operational decentralization arising from the compartmented nature of the terrain and potential for rapidly changing situations characterizes urban operations. The British, whose experience in Northern Ireland can be measured in decades rather than years, promote independent thought in subordinates and in fact have institutionalized means of capitalizing on such freedom. Post-patrol debriefings include an "honesty check"; patrol leaders review their original route overlays and "correct" them to reflect exactly what course they took. In so doing, they identify "shortcuts" and ground perhaps otherwise not covered. Alternatively, the process allows leaders to ascertain when a route might be used too frequently, thus increasing the force's vulnerability to terrorist booby-traps or ambushes. The honesty check is conducted in a completely nondisciplinary way. A unit leader is not punished because his unit took a shortcut. The point is to obtain as complete an intelligence picture as possible while maximizing both force effectiveness and security. These same procedures were employed (and adapted as necessary) during operations in Pristina in Kosovo. Such freedom to demonstrate initiative is not a universal norm in USMC units. Neither is allowing Marines to practice such unorthodox (by U.S. standards) procedures as donning another nation's military uniform to decrease the chances of reflecting undue U.S. interest in a particular area, target, or individual. The British used this technique in the Balkans to conduct drive-by or walk-by looks at targets (Kerkering, 2001).

Effective NCO leadership development, an area in which the Marine Corps reconnaissance community excels, is crucial to successful tactical urban reconnaissance. Those at the gunnery sergeant level are in particular trained to make decisions on their own, guided only by the stated mission and their understanding of the commander's intent. Yet proper preparation for decentralized operations has to go beyond simply encouraging and relying on increased initiative. Communications planning must include coverage of such contingencies as loss of radio or satellite contact. Alternative communications and "no comms" plans should be part of standard operating procedures (SOPs), practiced in training, and rehearsed to provide backup means of dealing with all likely contingencies (Ziegler, 2001).

ROEs should reflect a similar flexibility, both in their design and in the training conducted to ensure that they are properly understood. Rules of land warfare training need to focus on issues likely to cause reconnaissance elements the greatest problems in urban areas. For example, is a Marine acting properly if he incidentally kills a noncombatant child who is being used as a human shield by a legitimate target? Knowing the answer beforehand and rehearsing for such contingencies saves Marine lives and can reduce the subsequent psychological burden borne. ROE design and training also need to account for the sudden changes in tactical situations more frequently found in urban areas than elsewhere—e.g., when a civil support action suddenly transforms into a firefight or crowd control situation (Kerkering, 2001). Any but the most straightforward change in the rules of engagement should be accompanied by verbal explanations, an opportunity for Marines to ask for clarification, and a rehearsal or "war-gaming" session that would help confirm understanding.

Task Organization and Structure

> If you're working with your reconnaissance guys, you're working with a cut above. You can get them to do it [conduct one-man reconnaissance missions].
>
> LtGen Ernest C. Cheatham, USMC (Ret.)

Team size will be an issue of considerable importance during urban operations, whether at the STA team, division, or force reconnaissance level.[18] The already oft-mentioned high demand for large numbers of reconnaissance assets means that considerable pressure will mount to use smaller teams and thereby cover more NAI or complete other important tasks. Smaller teams have the additional benefit of attracting less attention and thereby increasing the chances of successful undetected infiltration. They also present a lower profile once in position. The negative aspects are significant, however. Smaller teams cannot carry as much equipment as those permitting a wider distribution of the load. Security, other than that directly related to concealment and detection, is a greater risk. Fewer eyes in fewer locations increases the chances of a surprise compromise, whether it is deliberate or accidental. No universally correct task organization exists for urban reconnaissance missions. Team composition (including equipment) will be very much a matter of situation-dependent and experience-based judgment. It is a topic

---

[18]The topic of reconnaissance team size was one that inspired considerable debate and a variety of opinions during RAND's interviews. For the many reasons cited in the ensuing discussion, views on what would be appropriate organizations for urban missions ranged from one-man to 12-man units.

currently without coverage in reconnaissance doctrine, a void that should be addressed immediately.[19]

Another organizational consideration pertains to the primary specialties that merit inclusion on urban reconnaissance teams. Few would argue that having a Marine able to speak the language of the indigenous population and enemy (or both, should they differ) is highly desirable. The benefits in obtaining valuable, highly perishable intelligence through field interviews of enemy prisoners or indigenous civilian personnel are readily apparent. The Army has made the innovative, even revolutionary, move of putting a counterintelligence expert (read "HUMINT specialist") in the Reconnaissance, Surveillance, and Target Acquisition (RSTA) squads of their IBCT.[20] This makes eminent sense given that the IBCT is viewed

---

[19]The U.S. Army recognizes the difficulties inherent in organizing for urban reconnaissance contingencies. Their FM 7-92, *The Infantry Reconnaissance Platoon and Squad (Airborne, Air Assault, Light Infantry)* (2001a, pp. 9–15), notes that "because of the unique complexities associated with the three-dimensional urban battlefield, the decision to enter an urban area may require task organizing of the reconnaissance platoon to handle the unique operational challenges. The platoon leader may be required to organize the platoon to allow for greater area coverage, or to put 'eyes on' multiple areas of interest simultaneously. He achieves this either by forming two [reconnaissance and surveillance] teams per squad—one two-man team and one three-man team—or, with the inclusion of one of the platoon radio operations, he can organize the platoon into eight two-man teams. When conducting offensive reconnaissance missions, the platoon will normally be organized into three reconnaissance squads, each with its own area, zone or sector of responsibility. When conducting security operations such as screening or guarding, the platoon leader may choose to organize the platoon into two-man teams (controlled by squad leaders) to cover all avenues in the dense urban terrain.

[20]The relevant Army manual reads that "one IBCT organization manned and equipped to conduct HUMINT operations is the reconnaissance troops of the RSTA squadron. Within this organization, it is the presence of the 97B, Counterintelligence (CI) agent that provides it with a marked increase in HUMINT-gathering capability over any previous battalion or brigade-level asset. The significance of these 97Bs at the tactical level cannot be overemphasized to those unfamiliar with the IBCT organization, since the CI asset organic to conventional units at the squad tactical level is a capability unique to the IBCT, and one that is atypical in the U.S. Army. . . . This radical addition of 97Bs to the unit's organic manning places over 36 CI agents forward deployed with the RSTA troops at the lowest tactical level. . . . The 97B with the reconnaissance squads passes HUMINT . . . information to the S2X cell collocated with brigade through separate source data channels. . . . This allows not only the RSTA squadron S2 to receive information directly from the deployed squad, but also the brigade's S2X, the primary staff agency responsible for all HUMINT collection operations in the brigade area of interest. . . . The second source of HUMINT-gathering capability within the IBCT is the brigade's HUMINT platoon. The HUMINT platoon is

primarily as a support and stability mission-oriented structure that can also conduct offensive and defensive combat actions. While it might well be argued that Marine Corps reconnaissance units are the reverse, the benefits of including greater HUMINT collection capability farther forward during urban operations seem irrefutable. The question is less one of whether such a capability is desirable than whether the benefits so derived would outweigh the cost of manpower trade-offs made in providing for the change in structure. The option is worth further study given the USMC's apparent acceptance of urban warfare as a component of many future operations.

Reports

> Nothing helps a fighting force more than correct information. Moreover, it should be in perfect order, and done well by capable personnel.
>
> Ernesto "Che" Guevara

Reporting procedures received considerable attention earlier in this analysis, but Marine doctrine-specified reports themselves also merit a review. Marine Corps Reference Publication (MCRP) 2-15.3B, *Reconnaissance Reports Guide,* has a significant number of reports that would be suitable for use during an urban reconnaissance mission. Those that appear in Table 3.2 are notable in this regard.[21] None of them specifically includes more than passing urban area considerations. Others included in the manual refer primarily to amphibious operations and are not recorded here.

---

an element of the brigade military intelligence company and is not a part of the RSTA squadron. . . . Many are surprised at how quickly the soldiers learned the hard skill traits, such as combat reconnaissance, required of the 19Ds, especially since the 97Bs receive limited training on these techniques at Fort Huachuca." (1-18, pp. 50, 51, and 64; p. 64 provides an organization chart.) A footnote describes the S2X section as "a unique addition to the IBCT O&O not previously seen in other units. The S2X is responsible for the planning, tracking, and execution of all HUMINT-gathering operations throughout the brigade's area of interest" (U.S. Army, 2001a, p. 51). For a discussion of USMC RSTA, see "A Concept for Marine Corps Reconnaisssance, Surveillance, and Target Acquisition" (MCWL, undated).

[21]The report categories are those of the authors. No such categorization exists in the doctrine.

Table 3.2

Marine Corps Reconnaissance Reports

Type of Report	Examples
Reports regarding the enemy situation	Contact Report Enemy Sighting Report (SPOTREP) Standard Shelling Report, Mortaring Report, or Bombing Report
Reports regarding friendly situation	Casualty Report Situation Report (SITREP)
Reports regarding terrain and/or construction	Bridge Report Railroad Reconnaissance Report (RAILREP) Tunnel Report Route and Road Report Drop-Zone Report Landing-Zone Report
Reports regarding communications	Frequency Interference Report and Worksheet (FIRREP) Meaconing, Intrusion, Jamming, Interference Report and Worksheet

Reconnaissance personnel would adapt the reports in Table 3.2 as necessary to account for mission-relevant, urban-specific factors. Nevertheless, additional report formats would be of value as the frequency of urban operations increases. These could include the following:

- Structure report/urban target folder: For use in describing a specific structure or complex of structures, the report would provide intelligence regarding access and egress (ground level, aerial, and subterranean), preferable approaches (based on fields of fire from the target and surrounding area), building material, special conditions (e.g., bars on windows, hardened building, security personnel on site), proximity to other relevant facilities (such proscribed targets as churches or schools, nearby police stations, military installations, or other sources of potential reaction forces), obstacles to artillery or air attack, and other universal and mission-specific elements.
- Noncombatant activity report: For use in identifying indigenous personnel's routines and behaviors of military relevance. It would include such factors as determination of periodic events

(market days, times of prayer, religious service hours and locations), individuals of notable influence, the nature and composition of interrelationships, the identity of those sympathetic to friendly force interests, and potential points of leverage (e.g., bribes, provision of medical care).

- Urban resource report: Akin to an engineer reconnaissance report (which identifies stockpiles of construction materials, transport, and the like), this would focus on mission-relevant assets found in the urban area of interest. In addition to those sites pertinent in any environment (fuel points, construction material yards, power plants, water sources, radio stations), these would encompass
  - sources of food for friendly force or noncombatants (large grocery stores, refrigerated storage facilities),
  - vehicle lots (new and used car sales, parking lots/garages in which keys are left with an attendant, valet parking), and
  - Communications nodes, such as telephone operating centers.
- Sustainment systems report: Logistics support during urban operations tends to be closer to front-line units. It can also be a lucrative target for conventional and irregular enemy forces. Reconnaissance units can abet optimal selection of locations for medical units, supply, fuel storage, and other capabilities by identifying structures that are in desirable sites, provide essential services, or seem suitable for the sustainment of combat operations. Examples include facilities providing potable water or power, ease of operation under blackout conditions, direct access and egress, locations for temporarily storing the bodies of noncombatants or those killed in action, and surroundings that facilitate rear area security. Precluding theft of supplies and denying noncombatant access to garbage and food waste is a notable concern during stability and support missions. It is also a concern that units must be aware of during combat contingencies.

It should be noted that only reports essential to mission accomplishment should be required of any unit. SOPs and mission guidance should ensure that this is the case. Minimizing the number of reports has the added advantage of reducing radio transmissions that can compromise reconnaissance operations or allowing consolidation into a limited number of reports submitted by burst transmission.

Personnel Policies

There was general agreement that Marine Corps readiness would be enhanced by the assignment of appropriately experienced warrant officers to selected operational and training positions. "Gunners" are assigned to infantry units but not reconnaissance organizations. Consideration ought to be given to authorizing a warrant officer billet in divisional battalions. Assignment of such reconnaissance gunners to training positions would enhance SOTG training and instruction at the reconnaissance and dive schools. The SOTG gunner could replace the currently authorized 8654 specialty slot given that MEU(SOC) missions no longer include airborne or dive insertions.[22]

Further Training Implications

Additional concerns have arisen with regard to current procedures used in preparing Marines for urban ground reconnaissance. Direct action tactics, techniques, and procedures dominate SOTG instruction for reconnaissance elements, as has been noted.[23] Modification would be necessary should an adjustment to the current force reconnaissance emphasis on direct action be forthcoming. Solutions to the many concerns already noted (e.g., infiltration, exfiltration, call for fire, laser designation, OP positioning) would require instruc-

---

[22]Observation made during 2nd Reconnaissance Battalion interviews. Officer end-strength issues affect adaptation of current warrant officer assignment policies. However, the repeated concerns from the field would seem to merit a review of standing allocation guidance.

[23]A considerable number of serving Marine reconnaissance personnel of all grades interviewed expressed the concern that SOTG reconnaissance training is badly outdated and fails to meet current mission requirements. The perceived overemphasis on marksmanship (thought to be important but considered taken to the extreme in SOTG preparation) partially explains this perspective, but the nearly complete lack of urban cultural awareness, reconnaissance technique, and urban training other than building takedown instruction exacerbates these attitudes.

tion and practice, the latter likely at a minimum including terrain walks in actual civilian urban areas. As has been the case with the many valuable lessons learned from MCWL urban experiments conducted to date, these training sessions might well serve as the sources for and stimulate adapting initial urban ground combat reconnaissance TTP.

The interviews conducted in support of this analysis resulted in a number of additional observations regarding what training would be necessary were reconnaissance units to be committed to urban environs:

- "Setting up vehicle-based surveillance is one idea; however, we have absolutely no training on it" (Root, 2000).
- "Learning to detect and install boobytraps would improve our security and survivability" (Root, 2000).
- "How does the Internet work in the country?" (Fitzgerald, 2001).
- "Training and intelligence need to identify what's [notable] in the infrastructure. What are the important parts of the power plant. Given two plants, which should I take out?" (Fitzgerald, 2001).
- "Target analysis training does a pretty good job, but they need to focus more on other than U.S. cities." Another shortcoming of such training: "It is available only once a year" (Fitzgerald, 2001).

These several observations point to categories of urban-specific training needs. TTP guiding urban infiltration, movement, and hide establishment will be considerably different than in other instances. Cities complicate reconnaissance, but they can also offer alternatives to normal ways of approaching tasks. Vehicles, both mobile and stationary, are commonplace in built-up areas; "hiding in plain sight" while driving around a metropolis may be an entirely feasible means of performing some types of operations. The abandoned vehicles or wrecks frequently found in developing nations may serve as hides, stashes, sensor or relay sites, or other assets.

Boobytrap and mine emplacement (lethal or nonlethal, for protection or warning, command-detonated or otherwise activated) requires additional thought in an urban environment. The proximity of innocent civilians and domestic animals may influence the selection of the system used and the way it is put in place. Neutralization or recovery of the boobytraps and mines will be desirable in many cases. The shielding from fragmentation, blast, and sound offered by different building materials should similarly influence where, how, and what capability is chosen. These or similar properties will also make a difference when a team employs communications systems, relays, and sensors.

Purpose-built U.S. urban training facilities provide a venue for practicing or testing low-level TTP. They are less effective for training units of more than perhaps company size or rehearsing functions that demand a broader operational perspective (such as reconnaissance). These shortcomings can in part be overcome by training on closed military installations and in actual U.S. towns and cities. However, few places in the United States adequately replicate the conditions found in a large, developing nation's urban entities. Undoubtedly, conducting terrain walks in American built-up areas has value, but training needs to address differences between domestic conditions and those Marines will likely confront overseas. Given the opportunity during "floats," commanders can conduct terrain walks and other familiarization in international urban areas.

Two final training notes. First, it is important that the USMC reconnaissance schools on each coast be given the flexibility to adapt their curriculum to the challenges of the theaters to which their Marines will deploy. However, training standards and procedures should be uniform across the Marine Corps. Therefore, those established and taught at both schools should be identical with location-specific alterations serving as adaptations of the common norm. Second, the effective compression of the battle space in urban areas increases the likelihood that covert, clandestine, and regular military forces areas of operations will overlap. Training and the doctrine that guides instruction need to address the command and control issues as well as other operational issues that this greater density of forces involves. The traditional solution of putting a no-fire or restricted-fire zone around "black" assets may be infeasible when "green" units must use streets or pass through buildings near the undercover force. These historical approaches designed for less-dense environments could be counterproductive—avoiding a given area might raise enemy suspicions regarding the protected region.

Physical and Social Infrastructure

> The center of gravity during operations may be the civilian inhabitants themselves.
>
> FM 3-20.96, RSTA Squadron

Some effects of urban infrastructure are fairly well known but still worthy of inclusion in plans and training (e.g., the negative impact that electrical switching yards have on tactical communications). The nuances inherent in others are less obvious. Infrastructure is potentially a tool of persuasion. Israeli denial of water and power to Yassar Arafat's headquarters in Ramallah in April 2002 is but one recent example. Effective use of this implement, and employment in such a manner as to achieve only the desired (and not counterproductive supplementary) ends, will require reconnaissance teams to identify critical nodes in infrastructure systems, critical components within those nodes, and how they all interact. (It is worthwhile to mention that the same is true of an urban area's social infrastructure. Appropriate cultural awareness training would set the preconditions for Marine reconnaissance personnel's ability to determine the significant points of influence in indigenous human relationships.)

Occasionally, complete devastation of a given part of an urban area is the best available course of action. If enemy forces are known to have concentrated in a limited number of locations, immediately destroying those selected structures or city blocks so occupied might be less costly in friendly force and noncombatant lives, and damage to civil infrastructure, than more deliberate and lengthy efforts that allow the foe to withdraw through successive positions. The decision to adopt such a course of action will lie with those senior to reconnaissance personnel, but the latter's ability to determine the limits of potential targets and the feasibility of the enemy's successful execution of retrograde movements could significantly impact that decision. Seeing the city as a system of components rather than a unitary whole facilitates envisioning and capitalizing on such alternative courses of action.

MARINE GROUND RECONNAISSANCE: TECHNOLOGY AND EQUIPMENT IN THE URBAN ENVIRONMENT

Introduction to the Fourth Theme

Fourth Theme. The urban environment makes special demands on equipment and technology. Testing in rural environments does not constitute testing for urban operations.

> Another thing that was sort of eerie . . . probably the most significant thing about urban combat was the noise, the noise ricocheting off buildings, and the dust. It was like you were always fighting with a smoke screen. . . . You could hear a guy down the block drop his weapon [at night]. The noise during the day compared to the total silence at night [was completely different]. . . . If you're going to perform reconnaissance at night you're going to have to have some sort of noise generation to cover the movement. . . . Noise echoes up and down hallways.
>
> LtGen Ernest C. Cheatham, USMC (Ret.)

> They won't give teams fragmentation grenades or Claymores, but such weapons are needed for contingencies in which a team is compromised. During one mission in Somalia, we inserted a team, and soon thereafter trucks with heavily armed Somalis got stuck right by them. They didn't have the firepower to protect themselves had they been compromised.
>
> 1st Force Reconnaissance Company Interview

The short-term focus, that of looking at improvements to Marine Corps urban ground reconnaissance capabilities that are feasible within the next five years, means that applicable technological improvements will be limited to already available or soon-to-be available capabilities. The following discussion notes observations of relevance in this immediate time frame and makes occasional mention of more time-distant developments that would be notable for their application to the missions under consideration.

Personal Armament and Basic Load

The short ranges that characterize most urban engagements and the constrained spaces through which reconnaissance Marines must move favor smaller and lighter weaponry. The size consideration goes beyond efforts to expedite movement and avoid compromise caused by metal striking urban surfaces. Shorter barrels and lighter weapons are better for close-quarters fighting; they can be handled with greater speed and brought to bear on an enemy more quickly. Carbines will therefore be in increased demand during urban fighting, but a requirement will remain for longer-barreled weapons and their greater long-range accuracy.

Training on these weapons should include rapid-reaction drills (also known as close-quarters combat or "quick fire" drills)—techniques for rapidly and accurately engaging a foe without bringing a weapon up to the shoulder for aiming. These drills need to include controlled automatic weapons firing. A high volume of initial firepower is no less beneficial in a built-up area than elsewhere, but greater accuracy is required to preclude fratricide or avoidable noncombatant casualties.

Regarding technical characteristics of such armament, ammunition selection (both that in the chamber and carried in the basic load) may differ for units fighting in cities. Concerns regarding minimum arming distances for such systems as the M203 and, in the near future, the Objective Crew-Served Weapon and Objective Individual Combat Weapon (OCSW and OICW, respectively) have been mentioned. If Marines carry M203s, conditions may dictate that they have a shotgun round chambered rather than a grenade because the number of overhead obstacles and minimum arming distance of the latter can result in grenades striking objects and rebounding into the friendly unit.[24] (It should be noted that the Vietnam-era shotgun round was found to have too few pellets and thus insufficient stopping power. The close-in blast and noise value of a good shotgun round during building interior engagements would be considerable. Future acquisition officials should review whether modification of the current round is called for.)

---

[24]The current grenade launcher will mount below the barrel on a M4 carbine. A standard six-man force reconnaissance team in 1st Force Reconnaissance generally carries M4s, two or three of which will be equipped with the launcher. One squad automatic weapon per team is authorized. Some units modify this weapon system by shortening the barrel, giving it a collapsible stock, and adding rails for the mounting of laser sites or light enhancement devices. The actual mix of weapons and components is mission dependent (Kerkering, 2001).

Selection of scopes must account for the shorter ranges that predominate during urban missions. Those with high-end magnification may have too limited a field of view to be of value for some urban scenarios (Ziegler, 2001, p. 5).

Systems for videotaping, directing a laser spot onto an individual's chest, or otherwise communicating that a person or group has the attention of a Marine force have proven valuable in operations with a stability component. While these procedures may not be directly applicable to combat situations, they could have a preventative or mitigating effect. However, in some cases, the ROE may prohibit the pointing of weapons at individuals. The NATO document *Improving Land Armaments: Lessons from the Balkans* (2001, paragraph 2.2.4.2) provides the following comment in this regard: "Systems need to be designed so that their sensors can be pointed selectively without aiming weapons at the same target. In many cases, sensors and video cameras were used successfully in crowd control situations and were more useful than guns. However, ROE often inhibit their use as they are tied to weapons, which, because of ROE, cannot be aimed or used." An alternative, of course, is to design ROE that do not deny use of such an asset.

While the standard basic load is probably sufficient for Marines with reconnaissance missions, elements on combat patrol or other direct action assignments may need to increase their size. The difficulty of creating premission stashes and obtaining rapid resupply during a firefight means that the standard quantities of magazines, grenades, and other weapons may not be adequate.

Vehicles

Mercedes Interim Fast Attack Vehicles are allocated on a three-per-team basis (thus two men per vehicle) in the 1st Force Reconnaissance Company. These vehicles have a spindle mount on which the Mk. 19, .50 caliber, or M240 weapon systems can be mounted. The M240 is not authorized for mounting at present, a seemingly unfortunate restriction given the lighter and smaller ammunition used by the gun. The vehicle was in part selected because up to two of them can fit in a CH-53 helicopter. (Some crews will allow only one.) One such vehicle fits in a CV-22 Osprey (Kerkering, 2001).

Vehicle insertion may be the best alternative during an urban contingency. The speed of movement, relative quiet, and increased load-bearing capacity of these vehicles provide advantages unequaled by foot or aerial entry into an operational area. Given that ground vehicles meet mission requirements, wheeled vehicles, such as those currently in the 1st Force inventory, are generally considered preferable to tracked alternatives. This is true of purpose-built combat vehicles as well as those modified from civilian chassis. Again, quoting from *Improving Land Armaments: Lessons from the Balkans* (2001, paragraphs 3.4.14 and 3.4.15): "Vehicles such as the BTR 80/A performed well as reconnaissance vehicles. Their mobility and reliability [make] vehicles such as the BTR 80/A ideally suited for reconnaissance missions. . . . Silence is very important. Tracked vehicles like the M113 family are too noisy and bulky for this type of mission. Depending upon mission and other planning considerations (e.g., mobility, survivability), consider use of adapted wheeled vehicles or sound-treated tracked variants (e.g., band track). Scout vehicles need to be mobile, stealthy, and flexible." Reconnaissance elements could use such vehicles to approach within foot-marching distance of their ultimate destinations; the vehicles could either be concealed or returned to origin points. Alternatively, further movement could employ bicycles (perhaps with trailers to increase carrying capacity) or, as one 2nd Reconnaissance Battalion Marine recommended, battery-driven (and therefore quiet) all-terrain vehicles (ATVs). An alternative would be to equip units with long-range ATVs, perhaps employing a hybrid model that uses battery power only for the ultimate silent approach (Schanz, 2001).

Unmanned Aerial Vehicles

The utility of UAVs is already widely recognized in the reconnaissance community. Marine units have long employed Pioneer or similar systems, and 2002 saw continued MCWL field experimental use of the Dragon Eye system. That there is benefit in augmenting planning and execution of ground reconnaissance with overhead efforts is undeniable, given appropriate mission profiles in which the UAVs will not breach operational security or compromise other efforts. Ideally, these systems will have a recording capability either through a live feed back to a recorder or an on-airframe system. Recording overflights reduces the need for repeated passes and thus the risk of shootdown. The tapes can also have subsequent enforcement, coercive, or negotiating value during stability missions.

Whether such aircraft ought to be organic to reconnaissance units is arguable. Some of those interviewed think such available UAV systems as the Dragon Eye are extremely valuable, an asset that the reconnaissance commander would likely want under his immediate control. Others consider the maintenance, training, and operator burdens too great, suggesting that it would be better were the asset assigned elsewhere. The potential timeliness penalties in this latter option are obvious.[25]

Sensors

Sensor development should provide Marine reconnaissance teams with portable, disposable, and camouflaged (or easily concealed) systems within the next several years.[26] Such capabilities, if able to meet the operational demands of the urban landscape, could increase team effectiveness and security. They could fundamentally alter reconnaissance TTP and team employment options—for example, two-man teams might be viable under conditions in which larger numbers would otherwise be necessary for security purposes. Depending on the effectiveness and reliability of such sensor systems, they could also reduce the number of manned reconnaissance commitments necessary to meet mission demands. Desirable capabilities would include vibratory, acoustic, and visual sensing and identification.

It is important to remember, however, that increasing information input from sensors would only further overload a system already unable to handle the volume of incoming information. If they are to be of value, technological advances that permit increased reporting

---

[25]1st Force Reconnaissance Company and 2nd Reconnaissance Battalion interviews.

[26]However, whether available sensors will be capable of providing the quality of input desired is questionable. Utility will be dramatically reduced unless new systems can detect, identify, communicate, and confirm readings with considerable reliability. Reliability depends on many factors, two of which are the ambient environment and expertise of system monitors. There can be a considerable period (in excess of 24 hours) between sensor emplacement and an operator gaining sufficient situational awareness to distinguish significant events from routine urban activity.

must be accompanied by analysis and dissemination capabilities that match collector enhancements.

Communications

Communications equipment was also found wanting. As noted, the H-250 radio handset was thought to be too loud, even in listening mode. Problems with LOS and fading led to requests for effective but lower-physical-signature urban antennas. Other long-standing requirements, such as secure, hands-free radio capability, need not be reiterated here.

Urban areas are communications-rich environments. Even cities in developing nations generally have hard-line and cell telephone systems. The ability to tap into either system for use in secure mode would be valuable and could help to overcome LOS problems in some instances (Rossignol, 2001).

Individual and Miscellaneous Equipment Needs

Urban environments make unique demands on some individual equipment. The noise made by hard-soled boots and the width of standard-issue Marine rucksacks (backpacks) were repeatedly cited during interviews as problems needing immediate attention. Both body armor and Kevlar helmets were considered essential for Marine urban ground reconnaissance. This is an interesting turn of events. The reconnaissance community has long been a group known for preferring soft caps to helmets and desiring to travel with the absolute minimum load feasible. Marines on both coasts called for an "urban drag bag" that would permit smaller reconnaissance teams to carry necessary equipment.

Coalition forces operating in the Balkans observed that laser range finders were valuable for observation and targeting purposes, particularly laser range finder binoculars that proved "to be a very good observation tool . . . particularly useful for targeting by [Mortar Fire Controller] and [Forward Observation Officer] parties" (NATO, 2001, paragraph 3.47).

The Marine Corps should consider acquiring earplugs that automatically seal or otherwise respond to excessive noise volume for use in subterranean passageways. This would address concerns regarding eardrum rupture when firing weapons in these enclosed spaces.

Individual thermal imagery equipment was highly sought after. Though the cost of the "Sophie" system is high ($58,270), the 20X magnification system is found to be very useful (Fitzgerald, 2001). It should be observed that thermal vision enhancement equipment is often not effective in fog. Such systems, whether for individual or sensor use, therefore need light enhancement, radar, or other complementary equipment items.

The following were all cited as desirable capabilities by one or more of the Marine reconnaissance personnel interviewed:

- Robots that could be used in sewer systems, ones capable of navigating in that environment and relaying their location (Kerkering, 2001). (Note that these could be used to create sketches or maps of subterranean systems before or in lieu of manned reconnaissance [should that alternative be necessary].)
- Power transformers—e.g., 220-volt to 110-volt (Jirka, 2001).
- Optics (fiber optic camera) to see into rooms and around corners (Morin, 2001).
- Better, lighter batteries with no hazardous materials—ones that will work in an airless environment (Morin, 2001).
- Remote listening capabilities, either sensors or other wireless means of allowing Marines to listen to conversations at a distance (Hinton, 2001; Hardy, 2001).[27] The monitored conversations could be transmitted to translators in a secure area (Jirka, 2001).
- Lighter, smaller, and more-durable communications, imagery, and computer equipment.

---

[27]RadioShack had a plastic child's toy as long as 15 years ago that looked like a pistol with a communications dish at the end of the barrel. When pointed at individuals having a conversation, the operator (wearing the earphones that came with the system) could hear what was being said at considerable distances. The technology behind football sideline sound amplification dishes is presumably similar.

Equipment and Systems

Two themes consistently ran through virtually any discussion of urban ground reconnaissance technologies. First, Marines repeatedly cited a desire to have equipment and systems important to their reconnaissance missions and known to work in open terrain tested in urban environments. Given the ever-increasing likelihood of urban operations, testing standards should include satisfactory urban performance. Second, either through individual commander initiative or USMC-wide acquisition, systems and equipment were brought on board in a manner that many thought ignored full life-cycle costs. Units therefore possessed equipment for which they felt Marines had inadequate training and funding for maintenance. Two comments are representative:

> "The Marine Corps is great about getting us a piece of gear when the money is there, but they never do the backdoor part of it [supporting and funding training, maintaining, and other life-cycle costs]."

> "The Marine Corps jumps all over in its purchasing. Every new system puts the user back at [square one]. There is a need to match purchase funds with funds for training, maintenance, and replacement."

As with all other components of Marine ground intelligence, technology issues require consideration from a systems perspective. As sensors, UAVs, and other capability enhancements reach the Marine in the field, frequency management will emerge as an increasing challenge. Frequency spectrum considerations should be considered to ensure long-term operational viability, including minimization of possible interference with mission-critical system elements already available and others under consideration.[28]

---

[28]For a fuller discussion of the frequency spectrum issue, see DoD (2002).

Chapter Four
CONCLUSION

As is more often than not the case with military capabilities, a commander confronting an urban combat operation will likely find himself with more reconnaissance tasks than assets to carry them out. Given that this commander has melded an organization capable of making the most of all elements of his intelligence collection system, he should be able to somewhat reduce the number of tasks assigned to ground reconnaissance units. Such wise use of the intelligence system also reduces the risk to which Marines in those units are exposed because reconnaissance obtained via unmanned or longer-distance means precludes the need to put individuals unnecessarily in harm's way. Ultimately, however, urban missions undertaken within the next five years will surely demand Marine boots on urban turf, for no other capability can see where they can see or go where they can go. Equally and surely, there will be more in the way of things to see and places to go than there are reconnaissance Marines to undertake the tasks.

The extent of these future shortfalls will in considerable part be a function of decisions made now. It is fortunate that many decisions can have immediate and significant effect. The role of STA teams, the nature of SOTG training, and the degree of flexibility designed into reconnaissance TTP are among those that can be altered in a matter of months after the provision of guidance so directing. Others, such as developing innovative urban infiltration techniques and testing them during exercises, experiments, and actual operations, will take more time, but developing an initial set of options for consideration should not be overly time consuming. A third class of decisions may extend beyond the grasp of immediate action. If

Marine Corps leadership seeks to significantly modify the character of force reconnaissance responsibilities but feels that firm joint community commitments exist, change (if desired) could be delayed by long negotiation and reallocation of strategic missions.

Whatever the outcome of decisions, be they maintenance of the status quo or dramatic revision, resultant urban combat ground reconnaissance TTP must be part of a training, planning, leadership, and operational execution system capable of continuous adaptation. Evolution of tactics in urban environments is potentially very rapid. The force that reacts quickly and effectively will have an edge over those that do not. A military capable of influencing an adversary's adaptation will have a further advantage in the ultimate struggle that is combat.

The mind-set of the current Marine Corps reconnaissance community and the service at large appears to be one conducive to favorably considering innovation and even radical change if the benefits merit. That it is time to initiate the development of reconnaissance TTP for urban ground combat operations has been recognized. Doing so from the perspective of perpetuating business as usual would corrupt the effort. The past offers much of value, but meeting the demands of the urban environment requires thinkers without too great a respect for "the way we've always done it."

Those fortunate enough to be given this important mission of creating a first-ever set of urban combat ground reconnaissance TTP should constantly remind themselves that Marine reconnaissance is both a system and a component of greater service, joint, multinational, and interagency systems. All of these systems change over time. The Marine that solves the problems of today without considering how environments, problems, and solutions will evolve by tomorrow has failed to best serve those at the cutting edge. Yesterday's urban combat was very manpower intensive. Today's commander has assets that allow him ever so slightly to reduce the number of his Marines that must burrow through buildings and meet the enemy in 25-meter engagements. The future will offer more in the way of such capabilities. Capitalizing on those to the extent possible ought to be a bull's-eye on which the Marine Corps lays its sights.

USMC URBAN GROUND RECONNAISSANCE SHORTFALLS

The following are the current Marine Corps urban ground reconnaissance shortfalls as listed in Chapter Two. They appear below as stand-alone elements without the explanatory material that accompanies them in the body of the report.

DOCTRINE

General

Formal, written urban combat reconnaissance doctrine is essentially nonexistent.

Intelligence collection in densely populated areas is more reliant on human intelligence (HUMINT) than is normally the case in other contingencies. Yet there is little guidance regarding how Marine commands should integrate this greater reliance on HUMINT into their collection and analyses processes.

Savvy employment of urban target systems analysis and urban intelligence preparation of the battlefield can enhance the value drawn from such HUMINT.

A need exists to further investigate the possibility that the complexity of urban areas may impose greater responsibility on Marine teams to provide analysis versus only reporting what is seen using the Size, Activity, Location, Unit, Time, and Equipment (SALUTE) report.

It is possible that urban reconnaissance teams must be better armed.

Counterreconnaissance guidance is lacking.

Specific Observations

There is a need to delineate Surveillance and Target Acquisition (STA), division reconnaissance, and force reconnaissance responsibilities relative to each other and to provide guidance with regard to their positioning that accounts for LOS, supporting fire, and communications limitations in the urban environment.

There is a similar lack of guidance on how to coordinate organic and external, in particular clandestine or "black" intelligence collection assets.

Prebriefings and immediate debriefings of civil affairs and medical personnel working with noncombatants should be incorporated into collection efforts, whether during Block 1, 2, or 3 missions.

Urban reconnaissance doctrine and training need to better identify requirements of other Marine units they are likely to support.

There is a lack of guidance regarding mission-relevant relationships between critical components of the civilian infrastructure.

Doctrinal guidance lacks information regarding conduct of subterranean reconnaissance.

Viable guidance is absent with regard to the insertion and extraction of reconnaissance elements.

Helicopter support operations are another area seen as requiring much more investigation.

Ground insertion **techniques have proven viable for allied forces, but the use of which has reportedly been denied by some Marine Corps leaders.**

Time factors for urban insertions and extractions are unknown and may vary from those in open terrain.

A need exists for planning and coordinating fire support plans to cover reconnaissance teams during reconnaissance missions and to minimize the number of changes to those plans during missions.

Similarly, urban escape and evasion (E&E) plans should be uniform and coordinated.

Marine air support for ground reconnaissance suffers from the same absence of doctrine and training opportunities as do the ground elements.

Combat reconnaissance elements may find themselves reconnoitering in support of multinational and NGOs/PVOs. No USMC doctrine exists that provides guidance with regard to proper execution of or training for these tasks.

Marine doctrine needs to discuss how reconnaissance assets can best aid leaders in shaping actions involving villages, towns, or cities.

The close proximity of STA teams and other reconnaissance assets to other friendly units during urban operations means that traditional reporting procedures may be inappropriate.

TRAINING

General

There is an outstanding and immediate need to develop a comprehensive and tiered approach to urban reconnaissance training that incorporates classroom instruction, drills, military training facilities, and actual urban areas.

The curriculum and standards for urban training should be consistent in reconnaissance schools and across units. Urban training packages that prepare units for the specific built-up areas in which they will operate during pending deployments should be tailored to meet local unit mission requirements.

There is a misunderstanding of weapons effectiveness in cities.

Specific Observations

Controlling urban fires is difficult.

Reconnaissance teams are at times not properly educated with regard to rules of engagement (ROEs) during deployments.

Cultural awareness and cultural intelligence training for urban reconnaissance personnel is an area requiring significant attention.

Other techniques thought to be of value but insufficiently covered in training are:

- Quiet and undetectable urban entry methods, such as picking locks and window latches or overcoming computer security systems.
- Gaining entry into and "hot wiring" vehicles for use when keys are unavailable.
- Better procedures for detecting, neutralizing, and installing boobytraps.

The lack of effective urban training involving units of greater than platoon size was considered a deficiency in USMC readiness.

Though communications, use of lasers, photography, and vision enhancement hardware has been improved in recent months, the lack of training that would permit testing these assets in urban environments leaves team members unsure of how built-up areas will influence technological performance during missions.

Urban environments impose special medical concerns for reconnaissance elements.

Marine reconnaissance training is currently too reliant on host nation support.

ORGANIZATIONAL, STRUCTURE, MANNING, AND PERSONNEL MANAGEMENT

General

Members of the reconnaissance community are unsure of what the optimum size is for urban reconnaissance teams.

It is necessary to "break the wall between the G2 and G3." Information of value to maneuver units at times never reached the elements most in need.

Specific Observations

There is a lack of specific information regarding urban infrastructure in mission areas and local national points of contact that can address specific related mission concerns.

A need exists to determine the echelon to which UAVs will be allocated and how they will be integrated into reconnaissance and intelligence dissemination systems.

Means to resupply Marines in hides, observation posts, or listening posts without compromising the position are currently lacking.

MATERIEL

General

There is a concern that too great a reliance on extant commercial off-the-shelf (COTS), military off-the-shelf, or brass board (in advanced concept, early development, or prototype form) products may fail to fully address identified needs in the interest of cost savings or immediacy of fielding.

Specific Observations

A need exists for acoustic or motion sensors that assist in detecting targets and potential threats in built-up areas.

Other wireless listening devices, including those that can amplify sounds over considerable distances or distinguish sounds through walls, would permit standoff collection of intelligence.

Design standards for equipment should consider the special demands urban environments put on end items.

A concern has arisen that new equipment purchases are too often not accompanied by the operator and maintenance training necessary to properly employ it.

Reliable communications and Global Positioning System (GPS) signals are areas of notable concern.

Communications compatibility even within the Marine Corps is terrible, much less with elements from other services.

There is a requirement for a stealthier means of monitoring radios.

The cumulative bulk of equipment was cited as a concern, one with special implications for urban operations.

There are several concerns regarding unmanned aerial vehicles.

BIBLIOGRAPHY

BOOKS

Adan, Avraham, *On the Banks of the Suez,* Novato, Calif.: Presidio Press, 1980.

Casper, Lawrence E., *Falcon Brigade: Combat and Command in Somalia and Haiti,* Boulder, Colo.: Lynne Rienner, 2001.

Geraghty, Tony, *The Irish War: The Military History of a Domestic Conflict,* London: HarperCollins, 1998.

Herzog, Chaim, *The Arab-Israeli Wars: War and Peace in the Middle East from the War of Independence to Lebanon,* London: Arms and Armour, 1985.

_____, *The War of Atonement,* London: Weidenfeld and Nicolson, 1975.

Nolan, Keith William, *The Battle for Saigon, Tet 1968,* New York: Pocket, 1996.

Norton, Bruce H., *Force Recon Diary, 1969–1970,* New York: Ivy, 1992.

von Freytag-Loringhoven, Major General Baron Hugo, *The Power of Personality in War,* translated by the Historical Section, U.S. Army War College, Harrisburg, Pa.: The Military Service Publishing Company, undated (written in the decade prior to World War I, the work was translated by Army War College personnel in 1938).

Warr, Nicholas, *Phase Line Green: The Battle for Hue, 1968,* New York: Ivy, 1999.

ARTICLES

Blevin, Don, "'Here Are Signs for the Wise'—Lessons of the Rhodesian War," *British Army Review,* December 1991, pp. 45–55.

Chuykov, V. I., "Tactics of Street Fighting: Military Lessons of Stalingrad," *The Cavalry Journal,* September–October 1943, pp. 58–62.

Cox, Matthew, "Snipers: Changing Mission, Changing Tactics," *Army Times,* November 26, 2001, pp. 18–19.

Crawley, Vince, "SpecOps Praised for Focus on 'Customers,'" *Army Times,* Vol. 62, March 25, 2002, pp. 18–19.

Ferry, Charles P., "Mogadishu, October 1993: A Company XO's Notes on Lessons Learned," *Infantry,* September/October 1994, pp. 23–31.

Gangle, Randolph A., "Training for Urban Operations in the 21st Century," *Marine Corps Gazette,* Vol. 85, July 2001, available at http://www.mca-marines.org/Gazette/Jul01Gangle.html.

Grau, Lester W., and Timothy L. Thomas, "Soft Log and Concrete Canyons: Russian Urban Combat Logistics in Grozny," *Marine Corps Gazette,* October 1999, pp. 67–75.

Grossman, Lev, "Welcome to the Snooper Bowl," *Time,* February 13, 2001, available at http://www.time.com/time/magazine/printout/0,8816,98003,00.html.

Jaffe, Greg, "Military Feels Bandwidth Squeeze as the Satellite Industry Sputters," *Wall Street Journal,* April 10, 2002, p. 1.

Kaplan, Karen, "The Sims Take on Al Qaeda," *Los Angeles Times,* November 2, 2001, available at http://www.latimes.com/templates/misc/printstory.jsp?slug= la%2D110201simosama.

Koch, Andrew, "Dragon Eye UAV Sets Sights on U.S. Marines," *Jane's Defence Weekly,* February 7, 2001.

Schenking, Scott, "Cavalry Operations in MOUT," *Armor,* Vol. 110, March–April 2001, pp. 15–17.

Shaver, Leslie, "Kick in the Door and Survive," *Marine Corps Times,* May 28, 2001, p. 12.

MANUALS AND REPORTS

Army Field Manual, Volume 2, Operations in Specific Environments, Part 5, *Urban Operations,* British Army, November 1999.

Army Field Manual, Volume IV, Operations in Special Environments, Part 5, *Operations in Built-Up Areas (OBUA),* British Army, 1998.

Army Transformation Taking Shape: Interim Brigade Combat Team (IBCT) Tactics, Techniques and Procedures, Fort Leavenworth, Kan.: Center for Army Lessons Learned, publication number 01-18, July 2001.

ATP-61, *Reconnaissance and Surveillance Support to Joint Operations,* North Atlantic Treaty Organization, November 1988.

Butler, Frank K., et al., *Tactical Management of Urban Warfare Casualties in Special Operations,* summary of 1998 meeting of the Special Operations Medical Association, October 25, 1999.

"Combat Operations After Action Report [for the period January 14–February 19, 1968]," Headquarters, 199th Infantry Brigade (Separate) (Light), April 15, 1968.

Edwards, Sean J. A., *Freeing Mercury's Wings: Improving Tactical Communications in Cities,* Santa Monica, Calif.: RAND, MR-1316-A, 2001.

Glenn, Russell W., et al., *Ready for Armageddon: Proceedings of the 2001 RAND Arroyo-U.S. Army ACTD-CETO-USMC Non-Lethal and Urban Operations Program Urban Operations Conference,* Santa Monica, Calif.: RAND, CF-179, 2002.

Improving Land Armaments: Lessons from the Balkans, Brussels, Belgium: NATO, November 2001.

"Individual Training Standards (ITS) System for Infantry (Enlisted), OCCFLD 03," USMC, April 5, 1999.

"Marine Corps Order 3120.9B. Subject: Policy for Marine Expeditionary Unit (Special Operations Capable) (MEU[SOC])," USMC, undated.

Marine Corps Reference Publication (MCRP) 2-15.3, *Ground Reconnaissance* (final preediting draft), USMC, March 28, 2000.

Marine Corps Reference Publication 2-15.3B, *Reconnaissance Reports Guide,* Washington, D.C.: Headquarters, USMC, April 21, 1998.

Marine Corps Warfighting Publication 2-15.3, *Ground Reconnaissance* (final preediting draft), Washington, D.C.: Headquarters, USMC, March 28, 2000.

Marine Corps Warfighting Publication 3-17.4, *Engineer Reconnaissance,* Washington, D.C.: Headquarters, USMC, 1999.

Marine Corps Warfighting Publication 3-35.3, *Military Operations on Urbanized Terrain (MOUT),* Washington, D.C.: Headquarters, USMC, 1998.

Matsumura, John, Randall Steeb, Randall G. Bowdish, Scott Eisenhard, Gail Halvorson, Thomas J. Herbert, Mark R. Lees, and John Pinder, *Rapid Force Projection Technologies: Assessing the Performance of Advanced Ground Sensors,* Santa Monica, Calif.: RAND, DB-262-A/OSD, 2000.

Project Metropolis: Military Operations on Urbanized Terrain (MOUT) Battalion Level Experiments, Experiment After Action Report, Quantico, Va.: MCWL, February 2001.

U.S. Army, *The Infantry Reconnaissance Platoon and Squad (Airborne, Air Assault, Light Infantry),* Field Manual 7-92, December 23, 1992; with Change 1, Chapter 9, "Urban Operations," Washington, D.C.: Headquarters, U.S. Army, 2001a.

_____, *RSTA Squadron,* FM 3-20.96 (second coordinating draft), 2001b.

"Urban Attacks Tactics, Techniques, and Procedures (TTPs)," MCWL, Urban Warrior Military Operations on Urbanized Terrain (MOUT) MCWL X-File 3-35.1, November 30, 1998.

Urban Generic Information Requirements Handbook (GIRH), Quantico, Va.: Marine Corps Intelligence Activity, December 1998.

INTERVIEWS

Allison, John, LtCol, USMC (Ret.), interview with Russell W. Glenn, Triangle, Va., October 18, 2001.

Avidor, Gideon, Brig. Gen., IDF (Ret.), interview with Russell W. Glenn, Santa Monica, Calif., April 21–22, 2001.

Campbell, Scott, Maj, USMC, interview with Russell W. Glenn, Camp Lejeune, N.C., August 30, 2001.

Ciancarelli, M., Capt, USMC, 2nd Reconnaissance Battalion, presentation at Urban Ground Reconnaissance Conference, San Diego, Calif., March 13, 2001.

Chalmers, Douglas, Major, British Army, interview with Russell W. Glenn, Santa Monica, Calif., April 20, 2001.

Cheatham, Ernest C., LtGen, USMC (Ret.), interview with Russell W. Glenn, Virginia Beach, Va., August 14, 2001.

Christmas, George R., LtGen, USMC (Ret.), interview with Russell W. Glenn, Triangle, Va., June 28, 2001.

Downs, Michael P., BGen, USMC (Ret.), interview with Russell W. Glenn, Quantico, Va., August, 14, 2001.

Eggleston, Corporal, USMC, interview with Russell W. Glenn, Camp Lejeune, N.C., August 30, 2001.

Evans, Jeff, Maj, USMC, interview with Russell W. Glenn, Camp Lejeune, N.C., August 30, 2001.

Evans, Robert, Lt Col, British Army, interview with Russell W. Glenn, Santa Monica, California, April 20, 2001.

Fitzgerald, Robert B., Master Chief, USN, 1st Force Reconnaissance Company, interview with Russell W. Glenn, Camp Pendleton, Calif., July 12, 2001.

Frost, William M., First Sergeant, USMC, 1st Force Reconnaissance Company, interview with Russell W. Glenn, Camp Pendleton, Calif., July 12, 2001.

Garrett, G. P., Col, USMC, interview with Russell W. Glenn, Arlington, Va., February 21, 2001.

Givens, MSgt, USMC, interview with Russell W. Glenn, Camp Lejeune, N.C., August 30, 2001.

Hardin, Milton, Gunner, interview with Russell W. Glenn, Camp Lejeune, N.C., August 30, 2001.

Hardy, Jeffrey C., Sergeant, USMC, 1st Reconnaissance Battalion, 1st Marine Division (Rein), interview with Russell W. Glenn, Camp Pendleton, Calif., July 13, 2001.

Hatcher, Timothy W., Gunnery Sergeant, USMC, interview with Russell W. Glenn, Camp Lejeune, N.C., August 30, 2001.

Hinton, Daniel P., Capt, USMC, 1st Reconnaissance Battalion, 1st Marine Division (Rein), interview with Russell W. Glenn, Camp Pendleton, Calif., July 13, 2001.

Hisdai, Yaakov, Col., IDF (Ret.), interview with Russell W. Glenn, Latrun, Israel, April 10, 2000.

Jirka, Patrick A., Staff Sergeant, USMC, 1st Reconnaissance Battalion, 1st Marine Division (Rein), interview with Russell W. Glenn, Camp Pendleton, Calif., July 13, 2001.

Karen, Arye, Brig. Gen., IDF (Ret.), interview with Russell W. Glenn, Ashkelon, Israel, February 17, 2002.

Kerkering, Richard T., Gunnery Sergeant, USMC, 1st Force Reconnaissance Company, interview with Russell W. Glenn, Camp Pendleton, Calif., July 12–13, 2001.

Lewis, Marshall, Gunnery Sergeant, USMC, interview with Russell W. Glenn, Camp Lejeune, N.C., August 30, 2001.

McCarthy, Robert, Maj, USMC, interview with Russell W. Glenn, Camp Lejeune, N.C., August 30, 2001.

McMillan, William M., HMC, USMC, 1st Reconnaissance Battalion, 1st Marine Division (Rein), interview with Russell W. Glenn, Camp Pendleton, Calif., July 13, 2001.

Mearkle, Curt, Gunnery Sergeant, USMC, interview with Russell W. Glenn, Camp Lejeune, N.C., August 30, 2001.

Members of BLUEFOR and OPFOR, 3rd Battalion, 4th Marines, interviews with Scott Gerwehr, Russell W. Glenn, and Jamison Jo Medby, RAND, George AFB, Calif., February 6–7, 2001.

Miller, Gregory D., Gunnery Sergeant, USMC, interview with Russell W. Glenn, Camp Lejeune, N.C., August 30, 2001.

Modrow, Harold E., LTC, USA, Special Assistant, U.S. Medical Materiel Development Activity, telephone interview with Russell W. Glenn, April 9, 2002.

Morin, Nicholas J., MSgt, USMC, 1st Reconnaissance Battalion, 1st Marine Division (Rein), interview with Russell W. Glenn, Camp Pendleton, Calif., July 13, 2001.

Paulovich, Michael J., LtCol, USMC, discussion with Russell W. Glenn, San Diego, Calif., March 13, 2001e.

_____, interview with Russell W. Glenn, Newport, R.I., October 9, 2001f.

Pope, Dan, Maj, USMC, interview with Russell W. Glenn, Charlottesville, Va., August 16, 2001.

Rossignol, Kris A., Staff Sergeant, USMC, 1st Reconnaissance Battalion, 1st Marine Division (Rein), interview with Russell W. Glenn, Camp Pendleton, Calif., July 13, 2001.

Rutan, James R., Master Sergeant, USMC, 1st Force Reconnaissance Company, interview with Russell W. Glenn, Camp Pendleton, Calif., July 12, 2001.

Schanz, William, Gunnery Sergeant, USMC, interview with Russell W. Glenn, Camp Lejeune, N.C., August 30, 2001.

Shamni, Gadi, Brig. Gen., IDF, interview with Russell W. Glenn, Adam Training Area, Israel, February 18, 2002.

Sward, Thomas B., Col, USMC, interview with Russell W. Glenn, Va., October 18, 2001.

Turner, Eric J., Capt, USMC, interview with Russell W. Glenn, Camp Pendleton, Calif., July 13, 2001.

Zaken, Nachum, Brig. Gen., IDF (Ret.), interview with Russell W. Glenn, Latrun, Israel, April 10, 2000.

MISCELLANEOUS

Department of Defense (DoD), "Spectrum: A Critical United States Defense Asset (Draft, March 25, 2002)," March 2002 (briefing).

Gangle, R., "13–14 Mar 01 Urban Ground Recon Conference Summary," undated.

Gangle, Randolph A., "Project Metropolis Company Level Experiment, Quick Look," e-mail to BGen William D. Catto, November 29, 2000.

Glenn, Russell W., "Notes from Trip to George AFB, Victorville, Calif., February 6–7, 2001," February 8, 2001.

_____, "Urban Ground Reconnaissance Conference, San Diego, NAB Coronado, Building 15, March 13–14, 2001," March 14, 2001 (notes).

"Ground Reconnaissance Conference, March 13–14, 2001," no author, no date noted.

Hatcher, Timothy W., Gunnery Sergeant, USMC, written notes, "Urban Reconnaissance Questions," August 30, 2001.

Jones, J. L., Gen, USMC, *Marine Corps Strategy 21,* Washington, D.C.: Headquarters, USMC, November 3, 2000.

"Lincolnia I (Draft Report)," no author noted, July 19, 2001.

Mangan, Bryan T., Capt, USMC, "One Intelligence Officer's View of Urban Reconnaissance," undated, but provided soon after February 2001 Project Metropolis experiment, George AFB, Calif. Captain Mangan was the Intelligence Officer, 3rd Battalion, 4th Marines, during the phase of the experiment conducted in early February 2001.

Marine Corps Warfighting Laboratory (MCWL), "A Concept for Marine Corps Reconnaissance, Surveillance, and Target Acquisition (RSTA)," paper available from MCWL, Quantico, Va., undated.

_____, "Current News: New Blood-Clotting Material May Revolutionize Combat First Aid, available at http://www.mcwl.quantico.usmc.mil/active.html, accessed July 16, 2002.

Paulovich, Michael J., LtCol, USMC, "Report of Urban Reconnaissance Exercise Conducted Project Metropolis, 19–27 Jan 2001," 1st USMC Reconnaissance Battalion, memorandum dated January 27, 2001a.

_____, e-mail to Russell W. Glenn, "RAND Urban Recon Project," June 18, 2001b.

_____, "The Challenge of Reconnaissance and Scouting in the Urban Environment," Naval War College paper, October 10, 2001c.

_____, e-mail to Russell W. Glenn, "Force Recon and STA Teams," November 13, 2001d.

Prescott, Jody, LTC, USA, e-mail to Russell W. Glenn, "Questions," April 10, 2002.

"Project Metropolis Phase 1 After Action Report—The Combined Arms Team in MOUT," MCWL, 2000.

Rafferty, Russell, e-mail to Russell W. Glenn, "Image Databases," February 13, 2001.

Root, K. R., "After-Action Comments for PROMET 01," 5th Platoon Commander, Bravo Company, 1st Reconnaissance Battalion, 1st Marine Division (Rein), FMF, ISMC, November 14, 2000.

Ziegler, Brian M., 1Lt (USMC), "Scout Sniper Platoon PROMET AAR & Urban Reconnaissance Suggestions" undated, but provided soon after February 2001 Project Metropolis experiment, George AFB, Calif. Lieutenant Ziegler was the Scout Sniper Platoon Commander for 3rd Battalion, 4th Marines, during the phase of the experiment conducted in early February 2001.